# ETHICS IN COMMUNITY MENTAL HEALTH CARE

# ETHICS IN COMMUNITY MENTAL HEALTH CARE

## COMMONPLACE CONCERNS

Edited by

PATRICIA BACKLAR

*Portland State University*
*Portland, Oregon*

and

DAVID L. CUTLER

*Oregon Health and Science Center*
*Portland, Oregon*

KLUWER ACADEMIC / PLENUM PUBLISHERS
New York Boston Dordrecht London Moscow

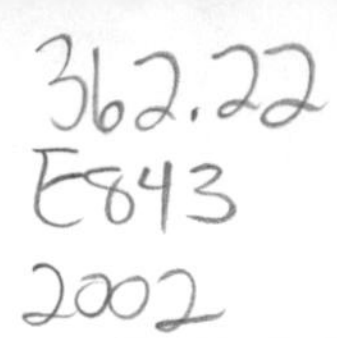


Library of Congress Cataloging-in-Publication Data

Ethics in community mental health care: commonplace concerns/edited by Patricia Backlar and David L. Cutler.
p. cm.
Includes bibliographical references and index.
ISBN 0-306-46704-6
1. Community mental health services—Moral and ethical aspects. 2. Psychiatric ethics. I. Backlar, Patricia, 1932– II. Cutler, David L.

RC455 .E85 2002
362.2′2—dc21

2001053923

ISBN 0-306-46704-6


233 Spring Street, New York, New York 10013

http://www.wkap.nl/

10 9 8 7 6 5 4 3 2 1

A C.I.P. record for this book is available from the Library of Congress



Printed in the United States of America

# CONTRIBUTORS

TANYA R. ANDERSON, M.D., is Associate Medical Director of Comprehensive Assessment and Response Training, University of Illinois at Chicago.

PATRICIA BACKLAR is Research Associate Professor of Bioethics in the Department of Philosophy, Portland State University, and Assistant Director of the Center for Ethics in Health Care, and Adjunct Assistant Professor in the Department of Psychiatry, Oregon Health & Science University.

CARL C. BELL, M.D., is President and CEO of the Community Mental Health Council and Foundation, Inc., and Director of Public and Community Psychiatry, and Clinical Professor of Psychiatry and Public Health, University of Illinois at Chicago.

MORRIS A. BLOUNT, M.D., is Psychiatrist at the Community Mental Health Council, Chicago, Illinois.

DOUGLAS BIGELOW, Ph.D., is Associate Professor in the Department of Psychiatry, Oregon Health & Science University.

RANDY BORUM, Psy.D., is Associate Professor in the Department of Mental Health, Law and Policy, University of South Tampa, Florida.

MARY ALICE BROWN, Ph.D., is Executive Director of Laurel Hill Center, Inc., Eugene, Oregon.

VALERIE COLLINS, M.D., is a Resident in Child Psychiatry, Oregon Health & Science University.

DAVID L. CUTLER, M.D., is Professor of Psychiatry, and Director of the Public Psychiatry Training Program, Oregon Health & Science University.

ROBERT E. DRAKE, M.D., Ph.D., is Professor of Psychiatry and Community and Family Medicine, Dartmouth Medical School.

GARY FIELD, Ph.D., Director of the Counseling and Treatment Program, Oregon Department of Directions.

FREDERICK J. FRESE, Ph.D., is Assistant Clinical Professor of Psychology in Psychiatry, Northeast Ohio Universities College of Medicine, and Vice President, Board of Directors, National Alliance for the Mentally Ill (NAMI)

JEFFREY GELLER, M.D., Ph.D., is Professor of Psychiatry, University of Massachusetts Medical Center.

CHARLES R. GOLDMAN, M.D., is Director of the Public Psychiatry Training Program, South Carolina Department of Mental Health, Division of Education, Training, and Research, and is Professor in the Department of Neuropsychiatry and Behavioral Science, University of South Carolina School of Medicine.

VIRGINIA ALDIGE' HIDAY, Ph.D., is Professor in the Department of Sociology and Anthropology, North Carolina State University.

COURTNEY JACKSON, is a medical student, Oregon Health & Science University.

HARRIET P. LEFLEY, Ph.D., is Professor of Psychiatry and Behavioral Sciences in the Department of Psychiatry, University of Miami School of Medicine.

JO MAHLER, M.S., is Senior Research Associate, Oregon Health & Science University.

BENTSON H. McFARLAND, M.D., Ph.D., is Professor of Psychiatry, Public Health, and Preventive Health in the Departments of Psychiatry, and Public Health and Preventive Medicine, Oregon Health & Science University.

CAROLYN C. MERCER, Ph.D., now deceased, was Policy Analyst at the New Hampshire-Dartmouth Psychiatric Research Center.

DOUGLAS L. NOORDSY, M.D., is Associate Professor of Psychiatry, Dartmouth Medical School, and Chief of Clinical Research in the Mental Health Center of Greater Massachusetts.

DAVID A. POLLACK, M.D., is Associate Professor of Psychiatry and Associate Director, Public Psychiatry Program, Oregon Health & Science University.

JEFFREY W. SWANSON, Ph.D., is Associate Professor in the Department of Psychiatry and Behavioral Sciences, Duke University Medical Center.

MARVIN S. SWARTZ, M.D., is Professor in the Department of Psychiatry and Behavioral Sciences, Duke University Medical Center.

H. RYAN WAGNER, Ph.D., is Assistant Professor in the Department of Psychiatry and Behavioral Sciences, Duke University Medical Center.

DEE WIRAK, M.S.W., is Co-Director, Rehabilitation Services, Laurel Hill Center, Inc., Eugene, Oregon.

# PREFACE

Why ethics and why now? Its certainly seems unusual for me to be concerned about ethics. It is an area that I always took for granted. I have always managed to see ethical problems as clear-cut and having simple, straightforward solutions. After all aren't there professional codes of ethics associated with each of the mental health disciplines to guide us? Don't we all know what these standards are?

The American Psychological Association has been working for years on drafting a new ethics Code. The final recommendations were to be made to the Executive Board of the American Psychological Association by the end of December 2000. This new version would need to be approved and that could take quite a while. In addition, the APA is considering no longer arbitrating ethics cases (it is time-consuming and burdensome financially to the organization). If that should become policy, then some in the organization feel there should be no updated Code produced as the APA will not be enforcing it.

Of course, a professional code of conduct is written from the perspective of a particular professional group, and does not take into account the needs of other stakeholders such as individual consumers or workers. In fact, most mental health workers are not even connected to an organization such as the APA so for most people such a relativistic account doesn't even matter.

With these thoughts in mind, I've begun to wonder about the validity of my assumptions regarding ethical practice in the mental health system. Three things have forced me to think like this (something I never intended to do). The first has been my involvement with the struggle for the rights of consumers to have a say in not only their treatment but also the systems that serve them. (A struggle I never consciously planned to be involved in.) The second is the revolutionary marketplace changes in the funding of public mental health programs. (A revolution that I still don't want to be involved in.) And the third is not a concept or a trend but a person. Patricia Backlar who is a health-care ethicist, author of the Family Face of Schizophrenia, a parent of a mentally ill person, and someone who refuses to face these dilemmas using denial as a defense strategy as I would prefer to do if she'd let me. On the contrary, she has persistently reminded me both formally and informally on a regular basis about all this ambiguity that is going on around us. What do you think of ...? or Have you an idea about how we are going to deal with ...? I finally decided that if I was going to have to face these things maybe the rest of you should also. And so I invited her to submit a series of papers regarding these various dilemmas to the Community Mental Health Journal. This was indeed a good decision judging from the cards and letters which have poured in. And so it occurred to both of us that maybe there is even a wider audience of earnest people who would choose to remain silent except for some prodding by our little chain letter of an idea. That is, what do some of the leaders in our profession, both consumers and providers, think about these things, and, specifically, what experiences have they had in dealing with them.

DAVID CUTLER, M.D.

# CONTENTS

PART II. PERSONAL BOUNDARIES

## Part VII. Conflicting Interests

# INTRODUCTION

PATRICIA BACKLAR

*Any conception of the ethical will include in some form a concern for people directly affected by one's actions, especially those to whom one owes special care ... .*
BERNARD WILLIAMS (1995)

*Psychiatry is inevitably entangled with our deepest moral concerns: what makes a person human, what it means to suffer, what it means to be a good and caring person. By the word "moral" here I do not mean a code of conduct of right behavior so much as our instinctive sense of what it means to be responsible ... .*
T.M. LUHRMAN (2001)

Socrates' question "how best to live?" is not an insignificant or esoteric query (Williams, 1985). Indeed, ethics is not a hifalutin subject. Many of us reflect on and ask a similar question. It is a question appropriate for all of us, whatever our abilities and disabilities.

Every day we are intimate with the experience of making choices. Many choices are mundane and straightforward, but some can be difficult and often we call such choices dilemmas. Dilemmas are endemic to being human. We call them *ethical* dilemmas when two or more moral interests are in conflict. Some people may believe that words like "ethics" and "morals" have a portentous rule laden quality. However, these words simply flag our effort to discern right from wrong, which in a turn, appear to spring from our human need to live in a community and our attempts to get along with each other. Thinking about `how best to live' is a common and enduring mortal endeavor. The examination of ethics is a deliberative and, at many times, an uncertain process. But above all else, the enquiry demands a communicative process (Backlar, 1995).

Our purpose, in writing and organizing this book, is to prompt—perhaps provoke—the reader to recognize, to reflect upon, to analyze, and to respond to the range of every day commonplace ethical concerns and dilemmas that arise in community mental healthcare practice with persons who suffer from severe and persistent mental disorders that may impede their ability to protect their own interests. The concerns that we have chosen to explore both encompass and interweave personal, social, and policy matters.

Ironically, at a time when most patients are living in the community—state mental hospitals having been downsized or closed—we find that ethics committees in the remaining mental hospitals are now burgeoning (Backlar & McFarland, 1994). Yet, for the majority of patients who now live in the community, there is no systematic process to address the ethical dilemmas that may occur. And this at a time when the mental health services for this heterogeneous population are in flux. Of course, the provision of these services in the community have always been in flux, have always been beset by insufficient resources, and have always struggled with difficulties that stem from the attempt to provide both medical services and social welfare services. Today, these long term challenges appear to be compounded by administrative disarray due to the melding of the public and private sectors: e.g., when a public mental health clinic is managed by a for-profit healthcare organization. Although this book does not directly address these enduring service provision crises, it is against this backdrop-the shifting systems for financing mental health care-that the chapters in this book have been collected.

The first chapter, ostensibly about ethical issues relevant to culturally diverse populations, actually is about ethical issues that clinicians confront in most of their everyday encounters with patients. Indeed—in a pluralistic society such as ours—Lefley considers that good practice involves a psychological component to cultural sensitivity that can enhance observational and therapeutic skills with individuals of all cultural backgrounds, including one's own. Similarly, in the second chapter by Brown and Wirik, which focuses on supported housing and rehabilitation services, the difficult ethical dilemmas—such as balancing rights and demands, and confidentiality versus preventing harm—are likely to be experienced by clinicians in a variety of settings.

The concept of personal boundaries, discussed in the next two chapters, are recognized as an abiding concern. The many-sided and divers nature of today's mental health services, as noted in Backlar's chapter three, may make healthcare professionals reconsider their traditionally held views about their roles, about their relationships with their patients, and about their understanding of customary boundary rules. More often than not, many providers no longer have the luxury of waiting for patients to visit the clinic, and some of those providers are likely to be peer counselors—persons who themselves are consumers of mental health services. In chapter four, Pollack identifies, in a series of case scenarios, the range and complexity of boundary concerns encountered in community practice. Analysis of the specifics in such real life scenarios can lead to effective policies and improved skills in administrative and clinical staff so that they become better equipped to respond to future ethical conflicts.

Violence perpetrated by persons with mental disorders is the topic addressed in chapters five and six. Backlar notes that in recent years society has shown an escalating interest in and concern about violence in the home. The major focus of that concern has been on child and spousal abuse, with a passing nod to elder abuse. Little or no attention has been paid to the despair and fear that some families with mentally ill relatives may experience in regard to threats of violence and actual violence. Furthermore, as Bell observes in chapter six, little has been written or noted about mental health clinicians' experiences with violence in the workplace—one issue that is rarely discussed is the personal safety of the clinician and how clinicians should respond to aggression.

In order to protect innocent individuals, there may be no way to avoid what many call controlling or coercive care for some persons with mental disorders who are unable, because of illness, to stop their violent behavior. In chapter seven, Noordsy, Mercer, and Drake's moral and clinical evaluation of involuntary interventions in the treatment of patients with dual disorders reveals that *how* the involuntary intervention is approached is just as critical as *whether* an involuntary intervention is used. Yet, as Hiday, Swartz, Swanson, Borum, and Wagner make clear in chapter eight, coercion itself—like outpatient commitment—cannot lead to positive outcomes for persons with severe mental illness if the mental health system does not have resources to give the treatment and services that are needed. Tragically, in the following chapter, "Why are severally mentally ill persons in jail and prison?" Cutler and colleagues confirm this observation. The beginning of the 21st century has been marked by an alarming migration of mentally ill persons into the criminal justice system. In truth, without housing and supports for living, it is impossible to coordinate and provide people with formal mental health services.

The next section discusses psychiatric advance directives. These types of documents were originally conceived as a self-advocacy tool for consumers to use in psychiatric emergencies. Over the years, debates have been waged about the feasibility of such psychiatric anticipatory planning. In chapter ten, Backlar describes research on Oregon's legal document, the "Declaration for Mental Health Treatment." And Geller, in chapter eleven, examines the experience with psychiatric healthcare proxies at a Massachusetts state hospital.

The conduct of psychiatric research is the topic discussed by Frese in the penultimate chapter. The author argues that consumers, individually and collectively, be given, as a matter of course, maximal input in all aspects of research that is done where consumer/patients are to be subjects.

In the final chapter, Goldman and Cutler speak to the increasing pharmaceutical industry influence on psychiatric research and education. They examine and consider several remedies for situations that create the potential for conflicts of interest and conclude that the burden of responsibility falls on professionals: the mental health professional has a fiduciary duty to place the interests of patients, research subjects, and science above personal benefit.

Most of the issues addressed in this book are not new. Indeed, many of us have become so accustomed in our daily lives to these concerns that we may give them short shrift. Yet, the first business of ethics is to discern what is an ethical dilemma. Dramatic moral concerns are easily recognized. However, what was originally a novel issue may become commonplace because we so often encounter the same kind of serious problem. For example, a young woman with schizophrenia refuses all treatment and care. Her delusions and hallucinations force her to suffer alone. Her parents, obliged to watch from a distance, are unable to offer her succor. They also suffer. And her concerned providers may be constrained—due to public policy, meager budgets, administrative obstacles, and even in some cases inadequate medical solutions—from giving her the care they know she needs. What we do in cases like this, how we are affected by our customs and our habits, and how we grapple with our decisionmaking is what we mean when we talk about ethics.

Problems and conflicts in any kind of system or policy emanate from personal positions. No solution is appropriate if it does not deal with the issues at the source (Nagel, 1991). For community mental health providers the issue at the source is the consumer. The consumers' good health and wellbeing is the *raison d'etre* for community mental health programs.

## REFERENCES

Backlar, P. (1995). Ethics in community care. *Community Mental Health Journal, 31*, 7–9.

Backlar, P., & McFarland, B. H., (1994). Ethics committees in state mental hospitals: A national survey. *Hospital and Community Psychiatry, 45*, 576–580.

Luhrman, T. M. (2001). *Of 2 Minds: The Growing Disorder in American Psychiatry*. New York: Alfred A. Knopf.

Nagel, T. (1991). *Equality and Partiality*. New York: Oxford University Press.

Williams, B. (1985). *Ethics and the Limits of Philosophy*. Cambridge, Massachusetts: Harvard University Press.

Williams, B. (1995). Moral Luck: a postscript. In W. Williams (Ed.), *Making Sense of Humanity: and Other Philosophical Papers* (pp. 241–247). Cambridge University Press.

# ETHICS IN COMMUNITY MENTAL HEALTH CARE

Part I

# COMMONPLACE ENCOUNTERS WITH CONSUMER/PATIENTS

CHAPTER 1

# ETHICAL ISSUES IN MENTAL HEALTH SERVICES FOR CULTURALLY DIVERSE COMMUNITIES

HARRIET P. LEFLEY

The question of cultural diversity raises a quintessentially ethical issue for all clinicians, that of fulfilling their promise to be a viable helping resource. It is a subject that calls for an educated awareness of how one applies professional training first to do no harm, and second, to deliver quality care and hopefully be optimally beneficial in serving the needs of one's patients.

Serious psychiatric disorders are now acknowledged to be panhuman and universally distributed, but they are assessed, treated, and experienced in a particular cultural milieu. The questions addressed in this chapter involve how best to serve individuals from cultural backgrounds that may be substantively different from one's own. At the very least, these issues involve conceptual clarity in communication, a common understanding of beliefs and values, and the ability to differentiate normative from maladaptive behaviors in cultural context. Functionally, cultural sensitivity may range from knowlege of variables that may affect a correct diagnosis to long-range treatment planning with the most potential for success. Understanding the cultural context will determine the relevance of a particular theoretical and therapeutic paradigm. Cultural sensitivity may also involve a recognition of differential service utilization patterns, a patient's potential compliance with medications, and type and scope of supportive resources available for community living.

There is also a psychological component to cultural sensitivity that can enhance observational and therapeutic skills with patients of all cultural backgrounds, including one's own. This is a mindset that incorporates an awareness of self, of one's own value system and possible biases in thinking, and how these may affect interaction with the individuals whom one is purporting to treat.

## DEFINING CULTURE

In its broadest sense, culture may be defined as a set of shared beliefs, values, behavioral norms, and practices that characterize a particular group of people who share a common identity and the symbolic meanings of a common language. In research across and within nation states, the definition has functionally been restricted to groups differentiated by race, ethnicity, or place of birth. In social science research today, cultural groups are also categorized along axes such as western versus nonwestern, or modern versus traditional. In recent years there has also been an increasing tendency to differentiate individualistic cultures, which give primacy to individual rights, from sociocentric or collectivist cultures, which focus on group loyalties and social role obligations (Triandis, 1995). This distinction generally overlaps with the respective world views of modern versus traditional cultures, and is related to the role of the patient in the family and to the family's role in caregiving. As we shall see, this cultural distinction is most likely to affect the interactions of families and mental health professionals and to inform their concept of confidentiality and the boundaries of the therapeutic alliance.

Social scientists also recognize subcultures based on social roles. These may include professions whose members have a shared belief system, a common symbolic idiom, and consensually standardized practices. On the basis of these criteria, it is fairly easy to characterize mental health professionals as a subculture (Lefley, 1998).

Some years ago, in developing a cross-cultural training institute for mental health professionals, I had done a review of the research literature and found that a wide range of issues affected the interactions of patients and clinicians from different cultural backgrounds. Most clinicians were middle-class whites from the majority culture, a group we indiscriminately identified as "Anglos". Most of the culturally diverse patients in these research studies were poor persons of color from minority groups. This large body of research indicated basic communication difficulties and bias in interviewing, both linguistic and psycholinguistic barriers in evaluating psychopathology, misinterpretation of psychodynamics, differential self-disclosure by patients, different expectancies of psychotherapy, therapists' false self-attributions of "color blindness" that interfered with authenticity and accurate evaluation, therapeutic advice that was counter to cultural mores, and failure to differentiate between adaptive and maladaptive behavior in cultural context. The research also indicated massive avoidance behavior on the part of the clinicians, as well as differential treatment of minority patients. Non-white, non-Anglo patients were less often accepted for psychotherapy, more often assigned to inexperienced therapists, and seen for shorter periods of time. They were significantly more likely to receive custodial care or drugs alone (Lefley, 1986b).

The avoidance behavior on the part of clinicians was matched by dropout and no-show rates of patients. The Seattle Project, a three-year study of nearly 14,000 patients in 17 community mental health centers in the Pacific Northwest, had earlier indicated that about 50% of all clients from ethnic minority groups (African-American, American Indian, Asian-American, and Mexican-American) failed to return after the first session, a significantly higher percentage than the 30% dropout rate for whites (Sue, 1977). Yamamoto et al. (1982) had noted that a higher percentage of Asian and Pacific Island patients were found to be psychotic and chronically mentally ill on the Psychiatric Schedule for Asian Americans, when compared with norms for non-Asian patients on corresponding psychiatric scales. Nevertheless, there was 60% less utilization of mental health services by Asian patients. This underutitlization pattern seems to have continued (Takeuchi, Mokuau, & Chun, 1992), although research by Okazaki (2000) found that Asian families delay treatment only when they feel high levels of shame and stigma. The literature indicates that Hispanics also tend to be underutilizers, while African-Americans vary, tending to be higher utilizers of outpatient public mental health facilities but to drop out earlier (Lefley, 1990).

One of the critiques of the older studies was that differences in treatment of minority groups failed to control for patients' socioeconomic status (SES). A more current analysis comparing treatment of African

American and European-American psychiatric inpatients found that when SES and diagnosis were controlled, many racial differences cited in earlier studies of psychotic patients were not statistically signficant. However, even with these controls, they did find racial differences related to the detection, phenomenonology, treatment, and course of psychotic disorders. Differences were also found in the diagnosis and management of substance abuse and personality disorders (Chung, Mahler, & Kakuma, 1995). Today the majority of persons with severe mental illness reside in the community. Barrio (2000) has analyzed the research and practice literature and questions the cultural relevance of community support programs. She suggests that mental health facilities incorporate cultural factors in psychosocial assessments, train staff in ethnographic interviewing, and use focus groups to understand the cultural backgrounds of the clients whom they serve.

But barriers to treatment are not unilteral. They may be superimposed by the system, but they are also self-imposed. In groups where major mental disorders have been historically stigmatized, persons who are highly symptomatic are often reluctant to use psychiatric services, and their families are unwilling to bring them. This has also been the case for ethnic groups who are generally unfamiliar with the services offered by the mental health system and have long tended to use other cultural resources. If remedies are required, the tendency is first to use pastoral or religious healing, or even the local pharmacy, to deal with symptoms. Ultimately, however, the literature suggests that patients who use religious healing systems will use the biomedical system as well. Research indicates that although alternative belief systems may be utilized for ordinary problems in living, people with major psychiatric disorders either use both systems concurrently or eventually rely on psychiatric care alone (Lefley, Sandoval, & Charles, 1998).

In this chapter, the ethics of healing subsumes a large array of relevant cultural variables. We consider some of the sociocultural stressors that are possible precipitants of decompensation in ethnic minority and immigrant groups, and the difficulties of disintentangling that which is purely ethnocultural from socioeconomic, minority, and immigrant status. In looking at ethnocultural variables, it is clear that people cannot benefit from our services unless they are willing and able to use them. Thus, we need to increase our understanding of service utilization patterns and barriers to treatment. We must also consider the extent to which people from culturally diverse groups tend to use alternative healing systems, professionals' attitudes toward these practices, and how these attitudes may deter or enhance effective treatment over the long range. We discuss diagnosis and treatment in cultural perspective. This includes the role of caregivers and family-professional relationships, and the cultural aspects of information sharing and confidentiality.

The values that inform our treatment modalities, and areas of conflict and concordance with the values and practices of many more traditional ethnic groups, may pose ethical dilemmas for western practitioners. Adherence to theoretical absolutes or to principles of cultural relativism may adversely affect practice. Of particular concern are treatment goals that may be unacceptable in one culture but adaptive in another. Our entire discussion of culture is framed within the ethical imperative of clinicians to do no harm and to be optimally beneficial to their patients. We end with research data suggesting that the outcome of cross-cultural training is not necessarily restricted to better treatment of patients from the groups under study. Our conclusion is that cultural sensitivity can result in heightened self-awareness and generalize to greater respect and understanding of all patients, regardless of ethnicity.

## CULTURE, ETHNICITY, AND SOCIETAL STRESSORS

Overall, when one speaks of cultural diversity it is hard to disentangle that which is purely ethnocultural from minority status, socioeconomic status, immigrant or refugee status, and linguistic or acculturative status, as well as the specific historical experiences of the group in question. All of these variables bring their own measures of stress which may affect behavior and symptomatology. A group's status and history within the larger society interface with those cultural beliefs and practices that may impinge on accurate diagnosis and treatment. Many of America's racial/ethnic minority groups are victims of centuries of economic and social oppression, of slavery, de jure and de facto segregation, enforced poverty, and deliberate policies of deculturation. Given this history, a substantial number of individuals have long developed adaptive behaviors, ranging from healthy anger to protective noncommunicativeness and suspicion, that may connote psychopathology to the inexperienced clinician.

Many members of our ethnically diverse populations are refugees or immigrants. Refugee status implies forced migration, typically an escape from social upheaval, frequently punctuated by transient resettlements before reaching the desired host country. There may be post-traumatic stress related to flight from a war-torn country, incarceration in refugee camps, loss of loved ones, and even starvation, rape, or torture. Even with voluntary migration, translocation from one culture to another, with attendant loss of family and old reference points, can be severely stressful for vulnerable individuals. In various countries, immigrants are known to have higher rates of hospital admissions for major psychiatric disorders (Leff, 1988).

Thus in addition to culture, there is a need to understand the common experiences of migration and loss and the stresses of acculturation, including almost ubiquitous intergenerational conflict. We need to understand how professional interventions can facilitate or disrupt the adaptation of patients and their families to a foreign land. The ambiguities of immigration status and citizenship status, in and of themselves, can be massive stressors. People may be legal immigrants or illegal undocumented aliens; they may be citizens or noncitizens, and each status brings with it a certain level of access or entitlement in the new society. Kinzie (1998) describes the effects of the so-called welfare reform bill that was signed into law in August, 1996. This law cut off supplemental security income (SSI) payments to disabled legal immigrants who had not been able to become citizens but suffer from physical or mental illness. He describes a patient suffering from PTSD, a war refugee, who plans to kill herself before the money is cut off. The anguish of refugees on the west coast was paralleled on the east coast, particularly in the Miami area, where a number of reactive sucides did in fact occur. Fortunately, provisions of that law have since been eased. But the massive stressor of immigration status, with all the forms and requirements for citizenship, including bureaucratic protocols for linguistic competency in English, continue to plague disabled individuals with horrendous pasts and uncertain futures.

## Utilization Patterns and Traditional Healing Systems

Whether they are migrants or indigenous, people from cultural backgrounds with their own healing traditions tend to be wary of mainstream mental health services. The conflicting etiological and treatment paradigms of western biomedicine and traditional healing systems often generate underutilization or misutilization of the treatment system, sometimes leading to tragic results (Fadiman, 1997). Hughes & Okpaku (1998) point out the cooperation that has emerged in recent years between Western medical practitioners and Navajo healers (called Singers). The Singers recognize that infectious diseases and those requiring surgery are better treated by Western medicine, but "Navajos also believe that Western treatment is only part of the restoration of harmony with the universe, and ritual treatment by a Singer is required for critical conditions" (p. 229). Clearly the same pattern obtains in the various supernational belief systems of Latin America and the Caribbean, as well as those of various Asian cultures.

Culturally sensitive practitioners should have some knowledge about the traditional or alternative healing systems in their locality, including mainstream religious healers to whom their patients may go for pastoral counseling. Many clinicians are uncomfortable with beliefs based on external locus of control and reliance on prayer for healing. Therapists can ususally accept the beliefs of patients who practice the major religions in the United States. They have far more difficulties with marginal belief systems and their healing methods, such as Mexican *curanderismo*, Puerto Rican *espiritismo*, Afrocuban *santeria*, Brazilian *umbanda*, Haitian *vodou*, Caribbean *obeah*, African-American rootwork, or American Indian medicine. All of these have explanatory models and correlative healing rituals for illnesses that often include anxiety, depressive and some dissociative states. Healers use herbs, perfumes and oils, exorcisms, trance possession, and other rituals that propitiate the powerful gods and remove curses inflicted by others.

A review of the literature as well as clinical experiences of those who work with healers suggests the following. First, alternative healers tend to be adept at differentiating major disorders, both physical and psychiatric, from those they can treat, and to make appropriate referrals to the biomedical system. Second, most patients who use traditional healing do so concurrently with medical and psychiatric care. Finally, clinical case studies suggest that in certain patients, rituals combined with psychiatric care may enhance response to psychotropic medications or improve compliance (Lefley, Sandoval, & Charles, 1998). Although much has been written about the non-Cartesian aspects of traditional healing, that is, the conception of mind and body as one, healers do differentiate between ills that are viewed as damage to the physical body and those that affect the emotions. As Hughes and Okpaku (1998) note, restoration of balance and harmony are an integral part of healing mental distress, and for believers—whether in western religions or other supernatural belief systems—this is a spiritual rather than a biomedical domain.

## DIAGNOSIS AND TREATMENT

Diagnostic objectivity has been a prevailing problem in the interactions of clinicians and patients from different cultural groups. Research has indicated that differences in the race and sex of patient and psychiatrist can influence diagnosis even when clearcut DSM criteria are used (Loring & Powell, 1988). Although we cannot attempt here to synthesize the massive literature on culture and psychopathology, a reference to current books on culture and clinical assessment can aid clinicians in making appropriate diagnoses. For example, diagnostic perspectives in treating African-American, Native-American, Asian-American, and Hispanic patients can be found in Mezzich, Kleinman, Fabrega, & Parron (1996). There are also sections on cultural aspects of the categorical disorders including organic and psychotic, mood and anxiety, somatiform and dissociative, childhood onset, and personality disorders, as well as culture-bound syndromes and multiaxial issues. In another cultural guide to clinical assessment, Tseng and Streltzer (1997) similarly deal with these discrete diagnostic categories as well as with violent and suicidal behavior and adolescent and geriatric psychopathology. Case vignettes and suggested clinical guidelines highlight the different cultural approaches.

Similar diagnostic categories, but with somewhat different approaches, are discussed in Gaw (1993). Okpaku (1998) gives an overview of clinical methods in transcultural psychiatry, with sections on treatment approaches in different cultural settings and a focus on women and children. In their review of the effectiveness of treatment of mental disorders worldwide, Sartorius (1993) and his associates from the World Health Organization discuss prevention, biological treatments, various types of psychotherapy (psychodynamic, cognitive, and behavioral), psychosocial and rehabilitative interventions, and the general influence of culture on treatment.

Race, ethnicity, and psychopharmacology are receiving increasing attention in the literature as it appears that Asians in particular, but other groups as well, may require different dosage levels of psychotropic medications. (Lin, Poland, & Nakaski, 1993). A prior review of this literature had indicated that findings in specific studies included extrapyramidical effects at lower dosage levels for Asian-Americans than for African or White Americans; lower dosage levels and a lower threshold for side-effects for antidepressants among Hispanics, and a better response by black patients to phenothiazines and tricyclic antidepressants (Lawson, 1986). Lin and his associates (1993) had noted that in addition to physical differences, cultural differences in personality styles and beliefs about drug effects may also affect response. They suggested that action-oriented patients with a need to control their environment, more typical of Anglo Americans, may require larger doses of sedating drugs than other patients. Among Hispanic patients, the confounding of culture-bound depressive somatization and antidepressant side effects was considered a factor in dosage response (Escobar & Tuason, 1980). More recently, Kinzie and Edeki (1998), looked at ethnic differences in drug metabolism based on a large number of studies from other countries, and found significant differences in poor metabolizers by ethnicity within race (Canadian, American, Estonian, Spanish and Swedish Caucasians; Chinese, Indonesian, Japanese, Korean, Vietnamese Asians, as well as African Americans and African American elderly). They point out that although biological factors such as body size and diet may affect response, non-biological factors may also have a role. Cultural factors may include the rate of placebo response in the population, the effects of the therapist's personality, and the patient's beliefs and expectations, which may be part of a culturally shared value system. Beliefs about whether medications are too "hot" or "cold" or too strong or weak, are found in many different cultures, and affect compliance with treatment regimens.

## Culture and Support Systems

In considering treatment of persons with chronic mental illness, it is incumbent on practitioners to know something about the support systems available in their communities; the resources such as churches and kinship networks, and the resources available to patients' caregivers as well. Worldwide, families are the major caregivers and support systems for patients (Leff, 1988). In the United States, patients from more traditional, collectivist cultures are not only significantly more likely to live with their families (Guarnaccia, 1998; Lin, Miller, & Poland, 1991; Lefley, 1996), but perhaps to be less conflicted about doing so—that is, they are less likely to feel devalued by a per ceived dependency status, and perhaps more willing to have their families engaged in a collaborative role in the treatment process (Lefley, 1998).

Marsh (1997) has discussed in considerable detail the ethical and legal aspects of the relationships of clinicians and the families of patients with serious mental illness. She talks about the perils of withholding information and support, and new models of translating ethical principles and theories into practice. The question of information-sharing with caregivers brings us to one of the most critical issues in psychiatric practice, the ethical parameters of confidentiality.

## Culture and Confidentiality

In the western psychiatric codes of ethics, the two greatest taboos for practitioners relate to sexual relations with patients, and to breaches of confidentialty, not necessarily in that order. Both are grave breaches of trust that contaminate the therapeutic alliance.

The issue of confidentiality strikes at the very basis of what is dearest in western individualistic cultures: the rights of patients to own their personal information. It is also underlies the basic premise of therapeutic progress. Separation and individuation are part and parcel not only of the psychodynamic paradigm, but of psychiatric rehabilitation as well. That is, there is an expected and sought for independence of the patient from his or her family of origin.

There is ample evidence from the research, however, that in more traditional, more collectivist cultures, it would be highly inappropriate not to involve the family in all matters concerning a patient's health and welfare. An article by two well known specialists in cross-cultural psychiatry points out the following:

> Notions regarding *confidentiality* differ across cultures. In some settings patients assume that any information conveyed to the clinician is a private and individual matter. Indeed, the laws and state apparatus may reinforce this belief ... In other societies the standard may be quite different. It may be assumed that anything conveyed to a clinician might be shared with the family, clan leader, or elders. In such settings, the unit of confidentiality may be the family rather than the individual. (Westermeyer and Janca, 1997, p. 302)

In our individualistic culture, this has clearly not been the general practice. Moreover, we are now in an era in which access to health and mental health records may be demanded by cost-saving strangers but withheld from those most intimately concerned with the patient's care and welfare. Petrila and Sadoff (1992) have suggested that the mental health professions should seriously re-examine the application of rigid confidentiality regulations, particularly when they compromise the ability of families to function effectively as caregivers. This is particularly salient in the light of culturally diverse patterns of home caregiving. Research suggests that African-American, Hispanic, and Asian families are much more likely than Anglo or European-American families to keep the patient living in the same household with them (Guarnaccia, 1998; Lefley, 1996; Lin, Miller, & Poland, 1991). There is a clear need for information-sharing on medications and illness management in order for families effectively to sustain this caregiving arrangement.

## ETHICS AND CULTURAL RELATIVISM

Modifying confidentiality to meet the practical needs of patient care in different ethnic groups is one example of cultural relativism. Probably the most critical issue in this discussion involves the application of one's personal value system within the context of a culturally flexible reality. Many of us subscribe to a model of cultural relativism that in itself is relativistic. That is, we yield to and accept customs that seem different but do not pose a value conflict with our own; we even admire some customs as superior to our own.

Other customs, however, are seen as defying some of our dearest convictions. Value conflict has always been a problem for psychotherapists, who are trained to be neutral and objective in their assessments and behaviors. Yet there are limits to stretching one's values, as many practitioners have found. How does an agnostic therapist deal with the soul-searching of a religious fundamentalist, when the patient has deep convictions that the therapist considers outlandish and socially damaging? How does the therapist avoid imposing his own world view while helping the patient come to terms with his own? How does a feminist therapist deal with the independence strivings of a woman who comes from a culture in which females are not only subjugated, but must be submissive in order to survive? Profound clinical depressions, and even psychotic episodes, are not uncommon among foreign female students, or wives of foreign graduate students, as they encounter the enhanced possibilities for women in the western world. Can a therapist in a university counseling center take responsibility for altering the life and future, and alienating the support system, of a woman who will return to her country and whom the therapist will never see again? These are not new problems in our field, but they are particularly salient today in an era of increased cultural exchange. In the following section, we consider some of the ethical dimensions of adhering to absolute and relativistic beliefs and values in patient care.

### THE TYRANNY OF ABSOLUTES

A professor is lecturing on boundary violations from a psychoanalytic perspective. He talks about the perils of accepting gifts from patients.

He also talks about violating spatial boundries, getting too close to patients, and particularly the taboo of touching.

A psychiatric resident protests that in Latin America, a physician *must* accept presents from his patients. To refuse a present is insulting. It demeans the gift and devalues the donor. All the Hispanic residents agree that refusal of a gift would be perceived as a profound personal rejection by the patient.

The professor insists that acceptance of a gift would be a boundary violation and hence damaging to the therapeutic alliance. This creates a double bind for the resident, who is trained to respect his elders and certainly his senior professors, but who knows in his very bones that refusal of a gift would be just like slapping his patient in the face. He cannot even tell the patient that the clinic rules forbid his accepting a gift. The patient sees him as the all-powerful doctor and hence the maker and breaker of rules. Besides, the resident sees no validity to the psychoanalytic dictum. In his culture, small gifts are pro forma and too insignificant to be interpreted as buying favor. They are symbols of gratitude, not bribes. There is a question of judgment here, but is it one of ethics as well? The resident suspects that by following the supervisor's dictates, he will hurt the patient's feelings and possibly induce abandonment of treatment. If the resident rejects his cultural knowledge in the interests of obeying or currying favor with a superior, is this an unethical act?

In a lecture on culture to a group of psychiatric residents, I talk about our Hispanic clinic and the compact between psychiatrist and patient. The Hispanic psychiatrists always end their sessions with male patients with a handshake and embrace. Dr. M. puts his arm around the patient and says, "Be sure to take your medicine, take it for me," and the patient answers, "I won't fail you, doctor."

A non-Hispanic female resident, Dr. A., shudders and says, "My supervisor told me *never* to touch a patient." We talk about the difference between touching paranoid patients and those who expect and want to be touched, about touching in inpatient and outpatient settings, and about touching a patient of the opposite sex. Dr. A. is not convinced that this rule can ever be breached, regardless of settings and circumstances. The discussion then becomes one of ensuring continuity of care. The psychodynamic paradigm is based on an ongoing therapeutic alliance. If adherence to this paradigm makes patients so uncomfortable it disrupts their attendance or worse, makes them drop out of treatment, there can be no therapeutic alliance. What is the ethical imperative—to follow the rules, or to keep the patient coming? In this case, does the end justify the means? What is the ethical imperative for the newly graduated psychiatrist—to follow a senior professor's directives, or to moderate them with his or her own cultural knowledge or even a gut reaction? When is it proper for clinicians to take the lead from their own convictions?

Finally, we ask, do clinicians have an ethical responsibility to follow current research, keep abreast of new knowledge, and question the applications of that which they have been taught? In the case of touching, for example, how valid is the therapeutic paradigm? Where is the empirical basis for all these essentially psychoanalytic conventions? Indeed, if adequately studied, they may turn out to be more countertherapeutic than therapeutic. Psychiatric knowledge is fluid and changing, and today's truisms may be tomorrow's travesties. Is it unethical to be dogmatic in any evolving field of knowledge?

## The Tyranny of Relativism

Being rigid about absolutes may be an inappropriate way of doing psychiatry, or any other helping discipline. On the other hand, what are the ethical dimensions of yielding to a strictly functional, relativistic view of therapeutics? How willing are we to countenance interventions that may be appropriate in a particular cultural setting, but would not be appropriate in our own? Most of us have a fairly clear idea of ideas and practices that are acceptable and unacceptable in terms of the moral code of our own culture. At what point, and for what purposes, are we willing to violate that boundary line? Certain questions may arise fairly often when therapists deal with patients from traditional cultures, and these issues may go far beyond those of countertransference and dealing with patients whose values are at odds with one's own. Therapists are usually expected to keep their own values private and to avoid imposing their own belief systems, but they are not expected to accept ideas that they find injurious or malevolent.

Therapists learn how to deal, for example, with a court-remanded skinhead adolescent spewing racial slurs and ethnic hatred. They certainly do not have to accept that adolescent's values, and in fact, will consider them pathological and strive to help the patient change them. But how far does a feminist clinician go when treating a woman whose religion dictates obedience to her husband or minister in all matters of importance to her? Can a therapist accept responsibility for helping patients change their values, divorce an abusive mate, or leave a stultifying community, when they lack the necessary skills for creating a new life?

Must a therapist keep silent about a practice that he considers immoral nepotism (such as placing an unqualified relative in an important job) but his patient's culture demands this as familial obligation? At this point the law decrees certain violations of the therapeutic contract, but only in cases of child abuse or threat to life. Under these conditions, practitioners' failure to speak would be illegal as well as unethical. But disciplinary codes rarely deal with the ethics of therapists' silence and its implicit message of neutrality, nor with the resolution of practitioners' own value conflicts and feelings of social responsibility.

## ETHICS AND OBLIGATIONS TO PATIENT, FAMILY AND SOCIETY

Western psychiatry has long been based on an ethic of primary obligation to the patient. Obligation to society supersedes this ethic only when the patient poses a clear and present danger to others. What if obligations to the patient involve imposing a potential long-term danger by deluding others as to the actual nature of the patient's illness? Consider the following type of intervention in China, arguably the very model of a collectivist culture.

A psychiatric research team wanted to initiate empirically validated interventions for families of schizophrenic patients. However, after surveying the western literature on various models of family psychoeducation and family therapy, they discarded them as inappropriate for China. Western models were all based on consensually accepted goals of making the individual less dependent on the family and an independently functioning member of society. There was also the presumption of an overall mental health care system that would provide ongoing services, a situation unavailable in China. Thus the family-based intervention in China became a very different type of psychiatric service. Here was the rationale.

> The ultimate social goal of family members of mentally ill individuals in China is to develop a sustainable family-based support system for the dysfunctional individual. The cultural and legal obligation of parents toward disabled children continue until the child is married, when the responsibility is transferred to the patient's spouse. ... The major goals for these parents are (a) to obtain stable employment for their child ... (b) to find a spouse for their child-often by minimizing the seriousness of the illness and offering a potential spouse (usually of lower social status) benefits

> such as housing ... and financial assistance; (c) to encourage the young couple to have their own child as soon as possible as this both secures the marriage and provides someone to care for the patient in his or her old age; and (d) to prevent divorce, by giving financial aid and practical help to the spouse.
>
> Once the patient's acute symptoms are under control, these issues become the major focus of the counselling sessions; the therapist assists patients and family members in their dealings with work sites, helps them negotiate the hurdles of marriage, educates them about the potential problems of pregnancy, and trains the patient to adapt to the demands of marriage and parenthood (Xiong et al., 1994, p. 240).

The prospect of "buying" a spouse for a mentally ill individual and minimizing the seriousness of the disability clearly poses ethical questions to western practitioners. They would also be ill-at-ease with a targeted goal of having children for the purpose of providing caregiving in the patient's old age. These children are of course, at higher genetic risk than the general population and might themselves require caregiving. If so, this would perpetuate a continuing sequence of disability, dissembling, and dependency.

The Chinese family intervention was based on education that would facilitate fulfilling a culturally acceptable goal. It was not education that may have modified or changed a goal that western practitioners might consider socially irresponsible or even unethical. On the face of it, this seems like a cultural paradox, Western practitioners in an individualistic culture would consider sacrificing the rights of the individual patient to have a spouse, children, and caregiving security for the good of unknown others in the society at large. The Chinese therapists, in a collectivist culture, participated in an effort to fulfill individual needs by potentially sacrificing the good of others. In providing therapeutic case management and counselling, they undoubtedly enhanced the coping strengths of the families in question. But from a western viewpoint, the clinician-reseachers also facilitated an activity that involved avowed dissembling and possible long-term injury to others.

What was the outcome? In a randomized controlled trial with three follow-ups, this family intervention in China had highly significant effects in lower proportion of rehospitalization, shorter duration of rehospitalization, longer duration of employment, and reduction of family burden (Xiong et al., 1994). It is also possible that this type of intervention, aimed at giving the patient a more normal life, might be concordant with the better prognosis for persons with schizophrenia in developing countries (Jablensky et al., 1991). I remember visiting a family in Bangalore, India. It was the home of a patient with chronic schizophrenia who had once been so sick he had a history of smearing feces on the walls. His family helped him find a wife by paying for her housing and university studies, and he held a job in his uncle's office for one hour a day. He did not live with his wife, but he visited and had lunch with her daily. The mother claimed the wife was quite happy with this arrangement. The young man was very pleasant and seemed quite stabilized during our visit.

One might question whether the appearance of normalcy, facilitated by a wife and children, can be compared with the actuality of a real job and independent living. And one may wonder how rewarding such an arranged marriage might be for the spouse. It is unclear, also, whether a woman with schizophrenia is as marriagable as a man, and whether the intervention proceeds along the same lines. But these are questions for our culture, not for theirs. According to the narratives of persons with schizophrenia, it is clear that in any culture, even the appearance of a normalized life is more desirable than one of disability and marginalty (Barham & Hayward, 1998).

## CONCLUSIONS

Numerous ethical issues are subsumed under the rubric of cultural diversity and any type of dependency status in our society, certainly including but not limited to severe and persistent mental illness. Backlar (1998) has cited the welfare reform law that restricted supplemental security income benefits to specific types of disabled noncitizen legal immigrants, and the destructive impact of anticipation of loss of benefits on these individuals. She states that "Whether young or old, poor or rich, weak or powerful, people's moral claims should not be reduced by their circumstances. Vulnerable persons who live in a community should be entitled to the same consideration and regard that the more influential members have guaranteed for themselves" (p. 128). This is an ethic that underlies our religious systems but not our political or economic systems. Western cultures have equated civil rights but not financial entitlements with the mere quality of being human. Yet the preservation of mental health is surely allied to the assurance of basic means for survival.

In the western world, cultural diversity is typically a euphemism for ethnic minority status, and ethnic minority status has historically been accompanied by social and economic deprivation. The two most underprivileged groups in the United States, African-Americans and American Indians, are disproporionately represented in psychiatric hospital admissions, a statistic variously attributed to greater existential stress, prejudice of police and psychiatric practitioners, and diagnostic error (Snowden & Cheung, 1990). If these interpretations are accurate, what are the ethical obligations of clinicians to address any or all of these putative stressors? Do mental health professionals have an ethical obligation to work on the political front to redress societal wrongs that affect their patient's lives? Do clinicians have an ethical mandate to learn about culturally different symptom manifestations in order to refine their diagnostic skills?

In this chapter we have talked about some of the issues that affect the diagnosis and treatment of culturally diverse groups, ethical aspects of the relationship of practitioners with patients' caregivers or support systems, and the ethics of cultural relativism. We began with a reference to a mindset of self-cultural awareness, which is the basis of all good clinical and rehabilitative interventions. The importance of this mindset in generalizing to all patients in one's caseload, regardless of ethnicity, was empirically demonstrated in research on the long-term effects of a national cross-cultural training institute for mental health professionals (Lefley, 1986a). Most of the participants worked in public sector community mental health centers or similar agencies serving deinstitutionalized patients with chronic mental illness. In their intensive exposure to cultural variables affecting mental health practice, the focus had been on African-American, Afro-Caribbean and Hispanic populations, reflecting the majority of the ethnically diverse patients served by the trainees.

The project's research findings showed significant improvement in clinicians' levels of knowledge, sensitivity and videotaped practice skills. Long term evaluation looked at the caseload statistics of both individual clinicians and their agencies on a pre-and post-training basis. The data showed significantly higher utilization of services by the indicated minority groups, and significantly reduced drop-out rates not only among the ethnic minority patients, but for all patients served by these agencies. Personal reports of the clinicians and clinical administrators attending the workshops attested to their feelings that the cultural sensitivity training had generalized to greater sensitivity to all patients served (Lefley, 1986a).

Sue and Zane (1987) have pointed out the obvious fact that patients come from many different ethnic backgrounds, and no practitioner can possibly attain relevant expertise in all these cultural groups. Their analysis of the cross-cultural clinical literature suggested that patients' perception of two interrelated processes, credibility and giving, are essential for therapeutic efficacy. Credibility means that patients must perceive the therapist as capable of helping them solve their problems, and there must also be a perception that something of value has actually been given. Sue and Zane (1987) point out that very small pieces of cultural knowledge that make the patient comfortable, like a pat on the back, can convey a problem-solving capability and an ability to give something viewed as beneficial.

Sue and Zane's hypothesis was supported empirically in the research findings of the cross-cultural training institute. In a time-blind evaluation of the videotapes of a brief therapeutic interview with a client from a different cultural/socioeconomic background before and after training, over 1000 student and clinician raters from three ethnic groups used a 20-item rating scale. There was a significant increase in perceived therapist efficacy following the training. A principal components factor analysis with varimax rotation found that a single factor accounted for 85% of the variance: credibility as helper, that is, perception of the therapist as someone with problem-solving abilities, someone to whom they would return for treatment (Lefley, 1986a, 1989).

Our assumption is that there is a relationship between the patient's perception of credibility and giving and the therapist's mindset of self-cultural awareness. This mindset goes well beyond the awareness that psychiatric residents are usually taught: that is, understanding one's own countertransference, the biases, emotional reactions, and general attitudinal response evoked by characteristics of the patient. It is an acknowledgement that, just like the patient, all clinicians bring to the therapeutic encounter a body of cognitions, beliefs, attitudes, and values in which they have been enculturated. This means they must look at themselves objectively not just as appliers of professional knowledge, but as participants in a joint interpretation of reality. It forces them to try to see through the patient's eyes. This is the beginning of respect for the other's viewpoint. And particularly for persons suffering from severe and persistent mental illness, the conveyance of respect, as we all should know, is the foundation of therapeutic effectiveness.

## REFERENCES

Backlar, P. (1998). Ethics in community mental health care: Justice for all? *Community Mental Health Journal*, *34*, 127–128.

Barham, P., & Hayward, R. (1998). In sickness and in health: Dilemmas of the person with severe mental illness. *Psychiatry, 61*, 163–170.

Barrio, C. (2000). The cultural relevance of community support programs. *Psychiatric Services, 51*, 879–884.

Chung, H., Mahler, J. C., & Kakuma, T. (1995). Racial differences in treatment of psychiatric inpatients. *Psychiatric Services, 46*, 586–591.

Escobar, J. I., & Tuason, V/B (1980). Antidepressant agents: A cross-cultural study. *Psychopharmacology Bulletin, 16*, 49–52.

Fadiman, A. (1997). *The spirit catches you and you fall down*. New York: Farrar, Straus, Giroux.

Gaw, A. C. (Ed.) (1993). *Culture, ethnicity, and mental illness*. Washington DC: American Psychiatric Press.

Guarnaccia, P. J. (1998). Multicultural experiences of family caregiving: A study of African American, European American, and Hispanic Ameican families. In H. P. Lefley (Ed.), *Familes coping with mental illness: The cultural context* (pp. 45–61), New Directions for Mental Health Services No. 77. San Francisco: Jossey-Bass.

Hughes, C. C., & Okpaku, S. O. (1998) Culture's role in clinical psychiatric assessment. In S. O. Okpaku (Ed.), *Clinical methods in transcultural psychiatry* (pp. 213–232). Washington DC: American Psychiatric Press.

Jablensky A., Sartorius, N., Ernberg, M., et al. Schizophrenia: manifestations, incidence, and course in different cultures. World Health Organization Ten Country Study. *Psychological Medcine*, monograph supplement 20, whole issue, 1991.

Kinzie, J. D. (1998). The question of welfare reform and refugee Fo's answer. *Community Mental Health Journal, 34*, 129–132.

Kinzie, J. D., & Edeki, T. (1998). Ethnicity and psychopharmacology: the experience of southeast Asians. In S. O. Okpaku (Ed.), *Clinical methods in transcultural psychiatry* (pp. 171–190). Washington DC: American Psychiatric Press.

Lawson, W. B. (1986). Racial and ethnic factors in psychiatric research. *Hospital & Community Psychiatry, 37*, 50–54.

Leff, J. (1988). *Psychiatry around the globe: A transcultural view*. 2nd ed. London: Gaskell.

Lefley, H. P. (1986a). Evaluating the effects of cross-cultural training: some research results. In H. P. Lefley, & P. B. Pedersen (Eds.). *Cross-cultural training for mental health professionals* (pp. 265–307). Springfield IL: Charles C. Thomas.

Lefley, H. P. (1986b). Why cross-cultural training? Applied issues in culture and mental health service delivery. In H. P. Lefley, & P. B. Pedersen (Eds.). *Cross-cultural training for mental health professionals* (pp. 11–44). Springfield IL: Charles C. Thomas.

Lefley, H. P. (1989). Empirical support for credibility and giving in cross-cultural psychotherapy. *American Psychologist, 44*, 1163.

Lefley, H. P. (1990). Culture and chronic mental illness. *Hospital & Community Psychiatry, 41*, 277–286.

Lefley, H. P. (1996). *Family caregiving in mental illness*. Thousand Oaks, CA: Sage.

Lefley, H. P. (Ed.). (1998). *Familes coping with mental illness: The cultural context*. New Directions for Mental Health Services No. 77. San Francisco: Jossey-Bass.

Lefley, H. P., Sandoval, M. C., & Charles. C. (1998). Traditional healing systems in a multicultural setting. In S. O. Okpaku (Ed.), *Clinical methods in transcultural psychiatry* (pp. 88–110). Washington DC: American Psychiatric Press.

Lin, K-M., Miller, M. H., Poland, R. E., et al. (1991). Ethnicity and family involvement in the treatment of schizophrenic patients. *Journal of Nervous and Mental Disease, 179*, 631–633.

Lin, K-M. Poland, R. E., & Nakaski, G. ( 1993). *Psychopharmacology and psychobiology of ethnicity*. Washington DC: American Psychiatric Press.

Loring, M., & Powell, B. (1988). Gender, race, and DSM-III: A study of the objectivity of psychiatric diagnostic behavior. *Journal of Health & Social Behavior, 29*, 1–22.

Marsh, D. T. (1997). Serious mental illness: ethical issues in working with families. In D. T. Marsh, & R. D. Magee (Eds.), *Ethical and legal issues in professional practice with families*. (pp. 217–237). New York: Wiley.

Mezzich, J. E., Kleinman, A., Fabrega, H., & Parron, D. L. (1996). *Culture and psychiatric diagnosis*. Washington DC: American Psychiatric Press.

Okazaki, S. (2000). Treatment delay among Asian-American patients with severe mental illness. *American Journal of Orthopsychiatry, 70*, 58–64.

Okpaku, S. O. (Ed.), (1998) *Clinical methods in transcultural psychiatry*. Washington DC: American Psychiatric Press.

Petrila, J. P., & Sadoff, R. (1992). Confidentiality and the family as caregiver. *Hospital & Community Psychiatry, 43*, 136–139.

Sartorius, N., De Girolamo, G., Andrews, G., German, G. A., & Eisenberg, L. (1993). *Treatment of mental disorders: A review of effectiveness*. Washington DC: American Psychiatric Press.

Snowden, L., & Cheung, F. (1990). Use of inpatient services by members of ethnic minority groups. *American Psychologist, 45*, 347–355.

Sue, S. (1977). Community mental health services to minority groups. *American Psychologist, 12*, 616–624.

Sue, S., & Zane, N. (1987). The role of culture and cultural techniques in psychotherapy: A critique and reformulation. *American Psychologist, 42*, 37–45.

Takeuchi, D. T., Mokuau, N., & Chun, C. A. (1992). Mental health services for Asians Americans and Pacific islanders. *The Journal of Mental Health Administration, 19*, 224–236.

Triandis, H. C. (1995). *Individualism and Collectivism*, San Francisco: Wesview.

Tseng, W.-S., & Streltzer, J. (1997). *Culture and psychopathology: A guide to clinical assessment*. New York: Brunner/Mazel.

Westermeyer, J., & Janca, A. (1997). Language, culture, and psychopathology: conceptual and methodological issues. *Transcultural Psychiatry, 34*, 291–311.

Xiong, W., Phillips, M. R., Hu. X., Wang, R., Dai, Q., Kleinman, J., & Kleinman, A. (1994). Family-based intervention for schziophrenic patients in China. *British Journal of Psychiatry, 165*, 239–247.

Yamamoto, J., Lam, J., Choi, W-I, Reece, S., Lo, S., Hahn, D. S., & Fairbanks, L. (1982). The psychiatric status schedule for Asian Americans. *American Journal of Psychiatry, 139*, 1181–1184.

CHAPTER 2

# ETHICAL DILEMMAS IN PROVIDING SUPPORTED HOUSING AND REHABILITATION SERVICES

MARY ALICE BROWN AND DEE WIRAK

## INTRODUCTION

Psychiatric rehabilitation services have evolved over the past three decades in response to deinstitutionalization and the limitations of traditional mental health treatment for people with severe and persistent mental illness. However, as these community-based, individually tailored services have become a major service modality, staff face ethical issues not previously encountered in more traditional office-based services.

Supported housing and other rehabilitation services are complementary additions to more traditional approaches that have often "over emphasized the medical aspects of the illness and under estimated the psychosocial aspects" (Brown, 1997, p. 147). Psychotropic medications helped consumers manage their symptoms but were rarely sufficient in helping them manage their lives. The challenges of community living were often overwhelming. Consumers struggled with the difficult tasks of locating housing, shopping for food, preparing meals and other activities of daily living, while learning to take medications as prescribed, and getting to their appointments at the mental health clinic. Treatment professionals were "focusing on the illness" while consumers were "focusing on their entire lives" (Ragins, 1994, p. 8).

Psychiatric rehabilitation (also referred to as psychosocial rehabilitation) services were developed to provide practical assistance to help individuals develop experience and the skills to compensate for the illness and to cope with the demands of everyday life. "At its most basic level, the process of psychiatric rehabilitation seeks to help persons with psychiatric disabilities determine their goals, plan what goals to work on first and how, and then develop the necessary skills and supports to achieve their goals" (Anthony, 1998, p. 80). Psychiatric rehabilitation programs help participants to regain their confidence and develop skills needed for working, socializing, and living in the community (Hughes, Woods, Brown, Spaniol, 1994). At the core of rehabilitation services is a set of values and principles that includes: active consumer involvement and empowerment; a holistic, biopsychosocial approach; emphasis on strengths and wellness; an emphasis on learning and doing together; the use of situational assessment and natural settings; social and community integration, and relationships with service providers that emphasize advocacy and partnership (Cook et al., 1996).

With the introduction of these services a new work force emerged. It was primarily a young, enthusiastic work force committed to the principles but "without the assistance of university-based educational programs that provided credentials for psychosocial rehabilitation" (Pernell-Arnold & Nesbit, 1990, p. 296). In the early years, program directors and supervisors relied upon on-the-job training supplemented by occasional relevant conferences and seminars to assist staff in developing and improving the skills they needed to be successful in their work with consumers. "These early professionals knew that creative and innovative strategies had to be continually designed to impact the varieties and complexities of disabilities caused by mental illness" (Pernell-Arnold & Nesbit, 1990, p. 296).

Innovation combined with a strong, consumer-oriented philosophy and non-traditionally delivered services meant that the lines between staff behavior considered "professional" and "unprofessional" were often blurred. The nature of the relationship was more collaborative and less prescriptive (Kisthardt, 1992). Traits such as professional distance and emotional detachment in the medical model were challenged. Ethical guidelines developed for staff providing mental health services in office settings, were not helpful to rehabilitation staff who did much of their work in the community—in consumers' homes, in grocery stores, restaurants and coffee shops, laundromats or consumers' workplaces. Rather than maintaining professional distance, rehabilitation staff were expected to be active in their consumers' lives, frequently working alongside them, doing activities together (Brown, Ridgway, Anthony, & Rogers, 1991). In addition to working with consumers, staff also needed to work in partnership with family members and others to help consumers access or develop the skills, resources or supports necessary to realize their goals. Staff approached their work with respect for the consumer and his/her family, and with enthusiasm and a "do whatever it takes" attitude.

This chapter focuses on some of the complex issues staff face when they are providing outreach services to people living in their own housing in the community. Supported housing staff (also referred to as outreach counselors) work with consumers individually in their apartments and neighborhoods, and assist them to develop the skills and supports they need to be successful. Supported housing services are organized around three central principles: (1) consumers choose their own living situations, (2) they live in normal, stable housing rather than in mental health programs, and (3) they have the services and supports required to maximize their opportunities for success over time (Carling, 1990). In addition to describing the complex ethical issues facing staff who provide these services, the chapter also identifies a process agencies can use to strengthen the organizational culture, reduce the likelihood of ethical challenge or crisis and increase consumer trust, confidence and competence.

## NON TRADITIONAL RELATIONSHIPS

### Challenging Multi-Dimensional Staff Roles

Each day rehabilitation staff provide a range of services to consumers—services individually tailored to meet consumers' real needs. For example, activities might include looking for an apartment, moving furniture, shopping for groceries, meeting to talk about goals, taking public transportation, teaching employment skills, working alongside a consumer at a work site, sharing a cup of coffee or sandwich, or engaging in social conversation. They may assist consumers in taking their medications as prescribed, help consumers manage their money and meet their basic needs, and access benefits such as food stamps, medical and dental care (Brown & Wheeler, 1990). "In these multi-dimensional relationships, the role of the staff may be unclear to both the staff and the consumer, and even vary from contact to contact" (Curtis & Hodge, 1994, p. 347).

Throughout these ordinary activities staff are expected to take the time to skillfully listen without being quick to judge or control (Laurel Hill Center, 1992). At times, staff are challenged by a consumer who is angry because he lacks money for a desired purchase or because he no longer wants to take his medications. In these situations it is easy for power struggles to occur and to lose sight of the consumer's goals and "use power as a tool to restrain behavior" rather than as a resource for "teaching self-management, risk-taking and decision-making skills" (Curtis & Hodge, 1994, p. 346). Staff need to be flexible, patient, creative, skillful and have good judgment. The relationship is fluid and changes depending on the goals being pursued, and is often filled with ambiguity. Staff may need to be medical consultant, coach, mentor, friend, adviser ... or even confessor to help a person recover (Ragins, 1994, p. 10). Staff work independently in the community, and it may be difficult to ask for help and for supervisors to know when help is needed.

## Creating Empowering Situations

The staff's ability to demonstrate understanding and concern for the consumer in ways that the consumer values is the heart of rehabilitation. It involves building trust, sometimes with individuals who don't trust anyone—particularly mental health staff. The process of rehabilitation is grounded in what a person wants to happen in his or her life. It involves learning what is important to consumers in their lives and how staff can help them achieve their goals. Mosher and Burti (1994) defined two of the most important questions that staff can pose in developing rapport as "What do you want?" and "How can I help you get it?" The identification of personal goals is both empowering to the individual and provides the context for rehabilitation interventions.

Each day as staff work with consumers, they face situations that require their independent judgment and skilled responses. By their responses, staff have the power to support or diminish a consumer's goal or dream. Outreach staff provide considerable support and assistance and can be a powerful influence is a consumer's life. Staff use power through their approaches in "defining problems, imparting information, influencing attitudes, managing resources, teaching skills, coaching behaviors, or overt coercion to effect compliance or containment" (Diamond, 1995, p. 18).

Supporting a consumer who has difficulty taking medication as prescribed can be viewed as one of collaboration or coercion. When consumers do not consider the prescribed medications as helpful, they are reluctant (or may refuse) to take them. When medications are viewed as a critical element of community tenure, staff may be required to deliver medications and observe the consumer taking them. One approach staff can use in empowering consumers and fulfilling job requirements is to work with the consumer to identify options and evaluate them. Feelings of control are directly related to the ability to make personal choices. For example, Hodge (1997) points out that the questions "What can I do to help you remember to take your medicine?" or "Which of these supports is most helpful?" are far removed from the statement "I'm going to have your medicine delivered every day." This subtle but important difference in the approach can be a major factor in compliance. "Compliance strategies must be developed in partnership with the person served" (p. 221). "When individuals with mental illness have the power to make decisions, they can be expected to take responsibility for them. Having control over aspects of one's life requires accountability for actions" (Hatfield, 1994, p. 7).

Staff can also assist a consumer in talking with his doctor about medication. If information about medication can be presented in terms of a practical benefit in areas of concern to the consumer such as help in controlling behavior that's leading to eviction from a desired apartment, rather than in medical terms such as decreasing paranoia or disorganization, it's more likely to be accepted (Diamond, 1983). The doctor may also discuss some of the options in type or dose of medication to provide the consumer with a greater sense of control. "Decisions regarding changes in dose or type of medication can become opportunities for collaboration between the individuals and their doctors" (Fisher, 1994, p. 14). The physician has knowledge about the range of medications available and their effectiveness in treating different symptoms. Yet, the consumer "is most familiar with his disease and has a valid point of view. We [consumers] are perfectly capable of studying, understanding, accepting, and dealing with our illness and its symptoms" (Leete, 1988, p. 51).

"One of the greatest challenges in mental health services is clarifying the use of power to influence or control the actions of another for the purposes of healing and safety" (Curtis & Diamond, 1997, p. 97). When decompensation occurs, staff frequently need to change their approach from making decisions *with* the consumer to making decisions *for* the consumer. It is far easier for staff to make good decisions for the consumer when a crisis occurs when staff know the consumer's wishes. An effective tool in helping consumers plan for a time when their decision-making capacity is impaired and to ensure that their desires are followed is the Advance Directive for Mental Health Treatment (Backlar et al., 1994; Backlar, 1995).

## Balancing Competing Rights and Demands

In addition to their responsibilities to consumers, staff may also have responsibilities to other entities—to the organization, to an employer, to the family, to the managed care organization, to the property manager, or to the larger community. Staff may be expected to help consumers take their medications, to convince them to go to medical and dental appointments when necessary, and to encourage them to participate in various rehabilitation activities. Staff may be involved in managing a consumer's money when the organization serves as payee for the consumer's financial benefits. At times staff may also struggle to balance consumer choice with medical directives. For example, what's the right thing to do when staff assists a consumer to shop for groceries and his food choices and preferences are in direct conflict with the doctor's orders? Or what should staff do when assisting a consumer who demands that cigarettes be included in her budget when she has severe respiratory problems?

Some of the most complicated ethical dilemmas involve a conflict between what is best for the person and what is good for the system or community (Curtis & Diamond, 1997). Supported housing programs often serve individuals who would have been confined to institutions only a few years ago but are now able to live in their own apartments with varying levels of staff support (Brown & Wheeler, 1990). Although these individuals have great needs, they may also refuse help, putting themselves at increased risk for victimization, rehospitalization and arrest. At times, their behaviors may raise concern in the community by those who find them bothersome or frightening.

The staff's role is to reach out and persuade consumers to accept services that they seriously need but resist. In addition to their "responsibility for establishing friendly and helpful relationships, [they also have] an economic function which involves assisting people in using the right amount of the most efficacious service" (Hodge, 1997, p. 215). For example, staff are expected to encourage consumers to telephone on-call staff when they're in crisis rather than going directly to the Emergency Room. The advent of managed care has made this job function even more important.

The central issue for staff is one of determining how strongly they should try to influence the behaviors, attitudes and beliefs of the consumers they serve. Curtis and Diamond (1997) point out that the methods used range from trying to influence and convince (approaches we use with friends and family) to coercion and overt control. In psychiatric rehabilitation, there is a strong value placed on respecting an individual's rights and humanness, and this may at times conflict with the program's responsibility to another agency such as the Social Security Administration. Some people served in supported housing programs are required to have a representative payee. Due to co-occurring drug and alcohol issues or a consumer's debts and difficulty budgeting and prioritizing spending to meet basic needs, payeeship may be at the consumer's request or may have been initiated by the referring hospital or a residential treatment program.

Money management is a needed service but holding or managing someone else's money creates a power differential that can be fraught with issues of control. The consumer's wishes and the responsibilities of being a payee (assuring that basic needs are met) can be in conflict or an ongoing source of consumer discontent. Regularly involving the consumer in budget planning, scrupulous record-keeping, regular contact with the Social Security office, and having someone other than the consumer's primary outreach counselor function as the money manager are very helpful strategies to prevent or reduce conflict. It is also helpful to include payeeship and money management as an area of quality assurance review.

In addition to these issues that staff may face on a daily basis, there are also more extreme situations that occur less frequently but definitely pose perplexing and challenging situations to staff and programs. The following section presents three situations that staff in our supported housing program recently encountered.

## CASE EXAMPLES—RECENT DILEMMAS

### Consumer Choice versus Concerns for Self-Harm

Carl, 39, had a long-standing history of homelessness and notoriety, and adamantly refused treatment for his schizophrenia. He was referred to the supported housing program after a serious medical illness had left him emaciated and physically unable to reside on the streets. Initially, he declined help. There was considerable pressure on the program to forgo the usual voluntary service provision in order to accommodate and meet this man's basic needs on his terms.

Staff worked to build rapport and assist Carl in maintaining an apartment in community housing provided by a local charity. Carl was willing to go out for coffee, benefited from staff assistance in getting his mail and paying his bills, and seemed to enjoy conversations (mostly ranting as he ate hamburgers). He made it clear that he did not want to discuss medication and dictated where, when and for how long he could be visited. For months he would not let staff into his apartment. He did not want to hear feedback about his hygiene, health and safety issues (bathing in motor oil while smoking, preserving excrement in jars and various other "experiments").

Carl did not want any other help. He was a very intelligent individual who was satisfied with his lifestyle and wanted to return to the streets as soon as his health would allow. Carl did not believe that he was mentally ill. Staff found it difficult to work with Carl and to accept his living conditions. Carl chose to live in filth and squalor, and his unusual "experiments" with feces were repugnant and cause for concern for his safety and that of the other tenants. He continually refused medical treatment even though his health was deteriorating. The only way staff could convince Carl to remove bags of garbage from his apartment was to negotiate the removal of a bag of garbage in exchange for a cup of coffee.

The outreach counselor continued to see him two to three times a week for two-and-a-half years until Carl had repeated angry outbursts demanding that staff stop bothering him, and spat on his outreach counselor. Staff reduced their visits to twice a month and Carl seemed to tolerate the new reduced schedule. When Carl finally became concerned about his health, staff assisted him in getting to the Emergency Room. Carl was given a fatal diagnosis and hospitalized for serious medical problems. Medications were prescribed for his health and edema. In spite of extensive efforts to secure an appropriate post-hospital placement, Carl was discharged to his home without follow-up due to his refusal of continued treatment. Staff talked with the authorities about whether involuntary commitment should be pursued, but there was disagreement about whether his medical condition met the standards for mental health commitment. Unable to get himself out of bed, believing he would recover from eating "tainted porcupine meat," Carl died six days after discharge.

Rehabilitation places a strong value on choice and self-determination and on preventing someone from engaging in activity that endangers his/her health or life. During the three years Carl was served, staff struggled to balance the value of self-determination (that Carl had very clearly defined) with the program's responsibility for safety (Carl's and his neighbors).

In Carl's situation, outreach staff had to continually assess how much they could negotiate for any change. Usually when a consumer is in danger of losing housing that he values and, staff can assist him and make it a collaborative effort. Carl remained ambivalent about living in his apartment versus living on the street, and clearly resisted "placement" in a medical setting. The fact that he died in his own place and not on the streets is remarkable in and of itself.

One of the most difficult times for a service provider is when a consumer's symptoms are increasing and he is obviously having a relapse, but he's unwilling to take medication or seek medical assistance. Waiting for and watching someone's behavior become endangering to the point of involuntary commitment is often grueling, especially knowing that the very process of decompensation causes harm. The longer, deeper and more frequently a person experiences psychosis, the more the brain is damaged. It is a challenge when staff are trying to assist someone who refuses treatment entirely and his/her life is filled with delusional beliefs and bizarre behavior.

Ethics implies choice. Ethical dilemmas arise when there is a conflict between competing values (Abramson, 1985). In Carl's situation, it was difficult for staff to determine how much to intrude in Carl's life. Even after Carl berated his outreach counselor and spat on him, staff continued to serve him (albeit less frequently) because Carl's health and life were seriously at risk.

Ethics provides broad moral standards derived from the principles of beneficence (do good), non-malfeasance (do no harm) and autonomy (respect self-determination) (Rosenbluth, Kleinman & Lowy, 1995). There aren't specific guidelines about when or how much staff should intrude on someone who wants to be left alone. Staff struggle with the issue of how much to reach out when "fired" by a consumer and the response varies from staff to staff and from one situation to another. "If we err in the direction of much intrusion, we risk alienating the consumer and the sins of commission. If we err in the direction of leaving alone, we risk neglect and the sins of omission" (Reamer, 1982, p. 268).

Richard Surles (1994) offered several factors for consideration when making these difficult decisions. These factors include: (1) the imminence and degree of danger; (2) the ability of the individual to understand the imminence and degree of danger; (3) whether the activity exposes others to danger; (4) the resources and limitations of the setting (e.g., hospital, community street, family home); and (5) the risk of intervening on long-term recovery outcome.

## CONFIDENTIALITY VERSUS SAFETY OF OTHERS

Supporting an individual through a bad decision and safeguarding the relationship when a person can no longer care for herself or an infant can be an arduous task. In the following situation, staff struggled to maintain the relationship with Mary even though there were times when staff couldn't support her actions.

> Mary is a Native American served by the supported housing staff. She was referred to the program because she was living a marginal existence. She eats poorly, smokes a couple of packs of cigarettes a day, abuses alcohol, methamphetamines, and marijuana. She is diagnosed as having a schizoaffective disorder, antisocial personality traits and hepatitis C. She makes decisions quickly, often without considering the consequences. Recently, she left town for several days without taking anything with her for her personal needs, and without arranging care for her three pet birds. Another time, she moved all of her essential items and half of her possessions out of town without planning for the next step. Staff assist her in managing her money by dispersing it to her throughout the week.
>
> Mary met Henry at the social program. Although they received training in safe sex and infectious disease control, it was apparently unheeded and Mary soon announced that she was pregnant. She was delighted by the news and planned to keep the baby. Staff assisted her in obtaining prenatal care and supervised living arrangements were also considered. Staff continued to work with her to improve her nutrition and decrease her smoking for the well being of the child. Mary didn't change these behaviors but she did stop taking her Clozaril.
>
> As Mary decompensated, she began to angrily hit her face and belly in response to her voices. Staff considered involuntary commitment but endangering the life of an unborn child was not considered grounds for commitment. Five months into her pregnancy she was hospitalized due to her verbal threats to others. Once she was stabilized on her medications, Mary improved and discharge was considered. At staff's suggestion, Mary was placed in a secure facility until delivery.

> Since this was Mary's first child, she didn't have a history with Protective Services and would retain custody of the infant until she demonstrated that she was unable to care for it. Once the baby was born, Native American tribal rights could make it more difficult to remove the child if serious problems occurred.
>
> The outreach counselor maintained her relationship with Mary throughout her pregnancy and helped to prepare her for the realities of raising a child. The more staff worked with her, the more concern staff felt about Mary's ability to care for the child. Mary signed releases and staff compiled a summary of her services, lifestyle and choices. The outreach counselor arranged with Protective Services to set up trials using a doll that Mary agreed to carry about and use in her skill training. Mary repeatedly misplaced the doll and a case was built. Staff talked with Mary about open adoption and the advantages this presented for Mary's continued carefree lifestyle. Eventually, Mary agreed.

The outreach counselor struggled to find an ethical path—considering Mary's needs and the needs of the unborn child. Although staff worked hard and hoped that Mary would follow the prenatal care regime and learn parenting skills, Mary was unwilling and unable to do either. The outreach counselor continued her supportive relationship with Mary and also notified Protective Services whose primary concern was the baby. Staff "may not always be able to help consumers get their way nor can they support and condone certain actions. However, they must be on the same side of the struggle as the person and must partner with them on many issues. [Staff] must express empathy and concern in a way that the person feels respected, supported and strengthened" (Hodge, 1997, p. 222).

How interesting that the antonym for dilemma is "solution." The process of finding a solution when emotionally torn by the options presented can be agonizing. The experiences that are the most challenging are the ones that push up against staff's own moral codes. In Mary's situation, the outreach counselor believed that Mary's baby must be protected if Mary was unwilling or unable to make the effort needed to care for it. Once she was able to find a resource that would focus on the welfare of the baby, the outreach counselor could focus on her relationship with Mary and be supportive.

An ethical argument could be made whether Mary fully understood how the information would be used when she signed the release. While staff did discuss it with her was there really informed consent? Did Mary realize that she could potentially lose her baby based on the information provided? In this specific situation, staff agonized about the welfare of the child as well as Mary's rights and interests, and felt it was important to inform Protective Services. If harm came to the child then great harm would come to Mary as well.

Another type of ethical dilemma occurs when staff serving one consumer have "sensitive" information about a second consumer because of the relationship between the two consumers. Protecting one consumer's health, safety, or interests while respecting the other's confidentiality on matters such as HIV, hepatitis C, criminal or predatory background status can be a quandary. The dilemma of confidentiality can frustrate not only those who want to know, but also those who do know but are obligated to safeguard the consumer's rights. There are times when staff wish that the parent, the unsuspecting partner, the community, or a legal authority could be told specific information. In rural areas, staff face this challenge with greater frequency and difficulty.

A recent situation presented this ethical dilemma.

> The program received a referral for Peter, 37, who was diagnosed with bipolar I, manic with psychotic features and antisocial personality traits. Peter had an extensive criminal history including assaults and kidnapping. The referral stated that Peter had a history of taking advantage of others who were vulnerable and that his last commitment had been extended when he assaulted one of the forensic unit staff.
>
> As staff got to know him, more charges become evident as well as his preoccupation with sex. Peter did not believe he had a mental illness, blamed others for his situation, made graphic comments about women and was also attracted to young boys. He was not mandated to, nor interested in receiving sexual offender treatment. He had never been charged for this offense and did not believe his behavior was a problem.
>
> Peter befriended Susie, another consumer in the program. They were attending church together and from his report it sounded as though he had his eye on a young boy there. Fortunately, Delores, another consumer who was familiar with Peter's history, attended this church and shared her concern with her counselor who was also on the team. The staff informed Delores that staff were bound by confidentiality laws, but that she was not. Delores feared retaliation. Several options were discussed including sending an anonymous letter to the minister. In her fear Delores did nothing other than to remain watchful and willing to report any additional concerns to staff since staff would be required to protect a known victim.

> Peter also spent time with Susie's family. He impressed Susie's father who recommended Peter for a part-time position at the community center. The community center did not do a background check. Peter was hired to provide security for youth basketball games and to escort young girls to their vehicles at night.

Peter has worked at the community center for two years and staff have continued to work with him. There is still concern about the potential harm he could cause to others although no incidents have occurred during his employment there. In this situation staff worked to safeguard the relationship and to continue providing skills training and support. The outreach counselor listened to Peter describe his contacts with people on the job without responding in a judgmental fashion, yet continued to remind him about legal requirements for age and consent.

There are times when staff are required by law to report specific information. In Peter's situation, if a potential victim had been identified, then disclosure to the authorities would have been permissible. By continuing to work with Peter, staff were able to monitor the situation and to assist him in learning to control his behaviors, learn new skills and develop a better understanding of himself and his life. Staff were able to provide timely, practical assistance to Peter because of an established relationship.

At times, staff may also need to warn or stop a consumer from revealing information or engaging in illegal activity in the presence of staff if the consumer doesn't want it reported to authorities. The staff's response depends upon the relationship between staff and consumer and the seriousness of the situation. Staff may warn a consumer that "If you tell me anymore about this… (or "if I see bruises…) I must report it."

To be congruent with personal ethics, staff must feel moral, honorable or right about their work. When a consumer's decision, lifestyle choices or actions feel indecent or unsafe to staff, ambivalence is a natural result. Staff's personal feelings can become an obstacle if the staff's beliefs or "gut level feelings" are in strong conflict with the consumer's decision. Staff can easily feel at an impasse when they can't rely on their feelings to guide them in facing their own dilemmas. Without personal convictions as a guide, staff may feel like they're entering into a blind alley or quicksand. When the consumer's decision is in conflict with staff's personal values, then sharing the dilemma with teammates and supervisor is a welcomed relief.

## CREATING AND MAINTAINING AN ETHICAL CULTURE

Staff supervision and team members' support are critical elements in effective rehabilitation services. Just as psychiatric rehabilitation services strive to empower consumers, rehabilitation centers strive to empower staff. "Creating an ethical culture means empowering people to do the right thing for the company, the customer and the community" (Bellingham and Cohen, 1990, p. 7). In the book Empowerment Takes More Than a Minute (Blanchard, Carlos & Randolph, 1996), the authors identify and describe empowerment as releasing the knowledge, experience and power that people already have inside them. They offer three keys to creating and maintaining an ethical culture. The three keys are share information with everyone, create autonomy through boundaries, and replace hierarchy with teams. Within an ethical culture, staff feel more empowered, the likelihood of ethical challenge or crisis can be reduced, and consumer trust and confidence are increased. However, taking these steps requires ongoing sensitivity and vigilance to issues that will arise when access to information is widened and roles are blurred.

### Share Information with Everyone

Psychiatric rehabilitation programs value partnerships and place a high priority on the involvement of staff and consumers. Active involvement depends on adequate and timely information. Staff need information about best practices in the field, about the mental health system, about the agency, about program services, and about working with consumers. They need to understand the culture of the agency and opportunities for involvement beyond their job responsibilities. Consumers need information about the agency's services, what they can expect from staff, their rights and responsibilities, ways in which they can provide feedback, and opportunities for greater involvement through participation in agency meetings, advisory committees and the organization's board of directors. Access to information is empowering to consumers, to staff and to the organization. However, with increased access to information and changing roles of participation at different levels of the organization, greater sensitivity and awareness is needed. Guidelines are needed to help consumers and staff change from more traditional consumer–staff relationships to working as colleagues with different experience and expertise. Guidelines are necessary but are not sufficient. It is the many opportunities for discussion and review that provide the vehicle for culture change.

## Step Two: Creating Boundaries for Autonomy

In step two, Blanchard points out that having clear boundaries creates a sense of purpose. The intention is not to inhibit staff with bureaucratic procedures but to offer clear guidelines that provide focus and a sense of security. A compelling vision and mission with an image of the organization that clarifies the purpose and values helps our staff see how their contributions make a difference. An organization's vision, mission and values provide the "big picture" that assists staff in translating the vision into goals and roles that define his/her part of the big picture.

When the agency philosophy, values, and mission are clearly stated and driven by valued consumer outcomes, both staff and consumers can be empowered to take more active roles. Staff expectations are clarified when the staff orientation and development process focuses on obtaining these outcomes and translates philosophy into guidelines for practice (Anthony, Cohen & Farkas, 1990). Consumers and staff learn what to expect and how to participate more actively in the rehabilitation process. Agency procedures outline the goal-setting process, delineating how services are to be delivered and documented to maximize consumer participation and describe progress toward goals. Program evaluation measures program goals and consumer outcomes. Quality assurance ensures that service delivery adheres to both internal value standards and the external standards imposed by oversight agencies (Brown, 1997).

Ethical guidelines are a tool to assist staff in creating boundaries for autonomy. Curtis defines "personal boundary" as the way "each person operationalizes ethics into his/her daily activities" and notes that staff establish these boundaries through "program expectation, example of others, past experience and personal comfort" (Curtis, 1992, p. 5).

The Code of Ethics developed by the International Association of Psychosocial Services (IAPSRS, 1996, 2001) describes the promotion of ethical behavior as:

(1) Practitioners are trained to recognize ethical issues and dilemmas
(2) Practitioners promote and participate in full discussion of potetial dilemmas and decision-making
(3) Practitioners consult with colleagues and supervisors regarding resolution of specific ethical dilemmas
(4) Practitioners take into account the IAPSRS Code of Ethics and perspectives of all stakeholders in deciding how to resolve or address dilemmas

## STEP THREE: REPLACING HIERARCHY WITH TEAMS

In traditional mental health settings, the model for supervision is usually one to one—supervisor to practitioner. If a therapist is troubled by an ethical issue, the supervisor is consulted. Within psychiatric rehabilitation settings where staff work in teams, discussing ethical concerns in team meetings is much more effective. Here staff can be encouraged to talk openly, sharing their experiences with team members. Having an opportunity to discuss their experiences in a group forum allows staff "to think through and contribute to the discussion and [it] presents the message that these dilemmas are not private matters relegated to discussion behind-closed-doors ... These are not special issues to discuss occasionally or only when there is a problem. They are part of day-to-day decision-making and must be surfaced often" (Curtis, 1992, p. 5). The discussions also help staff to learn parameters for independent judgment (Curtis & Hodge, 1990).

"When organizational cultures are not accepting and empowering of staff it is difficult for staff to develop accepting and empowering relationships with consumers" (Curtis & Hodge, 1994, p. 351). When staff feel empowered, they are more successful in empowering consumers. Sharing experiences (positive as well as difficult situations with consumers) help team members learn from each other and make better decisions. Ethical guidelines provide a framework for the way things should be.

In their multi-dimensional roles, working independently in the community, staff face challenging situations and struggle to find ethical solutions. Creating and maintaining an ethical culture in which concerns can be freely discussed helps to ensure that staff are empowered to do the right thing for consumers, the organization and the community.

## REFERENCES


Abramson, M. (1985). The autonomy—paternalism dilemma in social work practice. *Social Casework*, 66, 387–393

Anthony, W. A. (1998). Psychiatric rehabilitation technology: Operationalizing the "black box" of the psychiatric rehabilitation process. In P. W. Corrigan & D. F. Giffort (Eds.). *Building teams and programs for effective psychiatric rehabilitation, number 79*. San Francisco: Jossey-Bass Publishers.

Anthony, W. A., Cohen, M., & Farkas, M. (1990). *Psychiatric rehabilitation*. Boston: Center for Psychiatric Rehabilitation.

Backlar, P. (1995). The longing for order: Oregon's medical advance directive for mental health treatment. *Community Mental Health Journal*, *31*(2), 103–108.

Backlar, P., Asmann, B. D., Joondeph, R. C., Smith, G. et al. (1994). *Can I plan now for the mental health treatment I would want if I were in crisis? A guide to Oregon's declaration for mental health treatment.* State of Oregon: Office of Mental Health Services, Mental Health and Developmental Disability Services Division.

Bellingham, R., & Cohen, B. (1990). *Ethical leadership: A competitive edge.* Amherst: Human Resource Development Press.

Blanchard, K., Carolos, J. P., & Randolph, A. (1996). *Empowerment takes more than a minute.* San Francisco: Berrett-Koehler Publishers.

Brown, M. A. (1997). Community mental health programs: An administrator's viewpoint. In B. Blackwell (Ed.), *Therapeutic compliance and the therapeutic alliance,* (pp. 143–157) Amsterdam: Harwood Academic Publishers.

Brown, M. A., Ridgway, P., Anthony, W. A., & Rogers, E. S. (1991). Comparison of outcomes for clients seeking and assigned to supported housing services. *Hospital and Community Psychiatry, 42*(11), 1150–1153.

Brown, M. A., & Wheeler, T. (1990). Supported housing for the most disabled: Suggestions for providers. *Psychosocial Rehabilitation Journal, 13*(4), 59–68.

Carling, P. (1990). Major mental illness, housing, and supports: The promise of community integration. *American Psychologist, 45*(8), 969–975.

Cook, J. A., Pickett, S. A., Razzano, L., Fitzgibbon, G., Jonidas, J. A., & Cohler, J. J. (1996). Rehabilitation Services for Persons with Schizophrenia, *Psychiatric Annals, 26*(2), 97–109.

Curtis, L. C., & Hodge, M. (1995). Ethics and boundaries in community support services: New challenges. In L. I. Stein & E. J. Hollingsworth (Eds.). *Maturing mental health systems: New challenges and opportunities. New directions for mental health services* No. 66 (pp. 43–60). San Francisco: Jossey-Bass.

Curtis, L. C., & Hodge, M. (1994). Old standards, new dilemmas: Ethics & boundaries in community support services. (pp. 339–353). In L. Spaniol (Ed.). *Introduction to psychiatric rehabilitation.*

Curtis, L. C., & Diamond, R. J. (1997). Power and coercion in Mental Health Practice. In B. Blackwell (Ed.). *Therapeutic compliance and the therapeutic alliance* (pp. 97–122). Amsterdam: Harwood Academic Publishers.

Curtis, L. (1992). Boundaries and ethics in community services: Guidelines for decision making, *In Community, 5.*

Diamond, R. J. (1983). Enhancing medication use in schizophrenic patients. *Journal of Clinical Psychiatry, 44*(6), 7–14.

Diamond, R. J. (1995). Coercion in the community: Issues for mature treatment systems. In L. I. Stein & E. J. Hollingsworth (Eds.). *Maturing mental health systems: New challenges and opportunities. New directions for mental health services* No. 66. (pp. 3–18). San Francisco: Jossey-Bass.

Fisher, D. (1994). Hope, humanity and voice in recovery from psychiatric disability. *The Journal of the California Alliance for the Mentally Ill, 5*(3), 13–15.

Hatfield, A. B. (1994). Recovery from mental illness. *The Journal of the California Alliance for the Mentally Ill, 5*(3), 6–7.

Hodge, M. C. (1997). Case managers and compliance. In B. Blackwell (Ed.). *Therapeutic compliance and the therapeutic alliance* (pp. 211–224) Amsterdam: Harwood Academic Publishers.

Hughes, R., Woods, J., Brown, M. A., & Spaniol, L. (1994). Introduction. In L. Spaniol, M. A. Brown, L. Blankertz, D. J. Burnham, J. Dincin, K. Furlong-Norman, N. Nesbitt, P. Ottenstein, K. Prieve, I. Rutman, & A. Zipple (Eds.). *An introduction to psychiatric rehabilitation* (pp. 1–2). Boston: IAPSRS.

IAPSRS (1996, 2001). *Code of Ethics for Psychiatric Rehabilitation Practitioners*, 3–4.

Kisthardt, W. (1992). A strengths model of case management. In D. Saleeby (Ed.). *The strengths perspective in social work practice*. New York: Longman.

Laurel Hill Center (1992). *A demonstration of the possibilities*, 1–20.

Leete, E. (1988). A consumer perspective on psychosocial treatment. *Psychosocial Rehabilitation Journal*, *12*(2), 45–52.

Mosher, L. R., & Burti, L. (1994). Relationships in rehabilitation: When technology fails. In W. Anthony, & L. Spaniol (Eds.). *Readings in psychiatric rehabilitation* (pp. 162–171). Boston: Center for Psychiatric Rehabilitation.

Pernell-Arnold, A., & Nesbit, N. (1994). Training Psychiatric Rehabilitation Staff. In L. Spaniol, M. A. Brown, L. Blankertz, D. J. Burnham, J. Dincin, K. Furlong-Norman, N. Nesbitt, P. Ottenstein, K. Prieve, I. Rutman, & A. Zipple (Eds.). *An introduction to psychiatric rehabilitation* (pp. 296–297).

Ragins, M. (1994). Changing from a medical model to a psychosocial rehabilitatio model. *The Journal of the California Alliance of the Mentally Ill*, *5*(3), 8–10.

Reamer, F. G. (1982). Paternalism in social work. In F. Lowenberg & R. Dolgoff (Eds.). *Ethical decisions for social work practice*. Itasca, IL: F. E. Peacock. 254–268.

Surles, R. C. (1994). Free choice, informed choice and dangerous choices. In C. J. Sundram (Ed.). *Choice and responsibility: Legal and ethical dilemmas in services for persons with mental disabilities*. Albany, NY: New York State Commission on Quality of Care for the Mentally Disabled.

# Part II

# PERSONAL BOUNDARIES

CHAPTER 3

# ROLES, RELATIONSHIPS, AND RULES: BOUNDARY CONCERNS

PATRICIA BACKLAR

*My apple trees will never get across*
*And eat the cones under his pines, I tell him.*
*He only says, "Good fences make good neighbors."*
—ROBERT FROST, *Mending Wall*

Whenever someone mentions the word "boundaries," do you (like a slightly rebellious teenager) spontaneously begin to hum the words "don't fence me in" under your breath? In the words of the poet Robert Frost, "something there is that doesn't love a wall." Yet, recognizing and responding to boundaries (both spoken and unspoken) in our relationships with others seems basic to the development of our socialization skills. From our earliest years we experience the setting of limits. Most of us discover that there is no person with whom we can afford not to be circumspect. The concept of boundaries appears to be a subject in which we all are experts. Nevertheless, mental health providers of community support services are likely to acknowledge that it is hard to arrive at a common accord about fixed rules and guidelines in regard to some relationship boundaries (Backlar, 1996a; Gutheil & Gabbard, 1998).

Typically, the provider is thrust into the more powerful position in the provider/client dyad. The provider is likely to be responsible for the care of a client who, because of illness, anxiety, and a lack of information may be in a vulnerable state (Hellman, 1995). In light of this fiduciary relationship, a provider's familiarity with relationship boundary concerns and understanding of the relevance of limit-setting, appears to be necessary for the adequate performance of the professional role. Professional training teaches an appreciation of professional responsibilities within a trust relationship. Furthermore, some professional codes of ethics may specify rules that serve to protect a vulnerable client—in regard to the preservation of a professional distance—by proscribing the provision of health services where dual relationships may exist as with family members, sexual partners, friends, employees, employers, and other social or business associates (Gabbard & Nadelson, 1995).

However, negotiating with boundary issues in your own practice or explaining it to others may be far more confusing than you may anticipate (Gutheil & Gabbard, 1993). Indeed, additional obstacles may crop up when you find that your own professional code may be somewhat at odds with the codes of your colleagues in aligned mental health professions (Brownlee, 1996). At the outset boundary rules may appear quite obvious and uncomplicated, e.g., the prohibition of sexual contact between the healthcare provider and client. But the literature, and our own experience, confirms that even this may be a complex many-sided matter (Appelbaum, 1990; Appelbaum & Jorgenson, 1991; Gabbard, 2000; Gutheil & Gabbard, 1992). What really gives the concept of boundary issues its protean slipperiness—its variable shapes and forms—is not only the changes that may occur because of the shifting needs of discrete individuals in dissimilar circumstances, diverse relationships, or various locations, but also the changes that have been taking place in the very practice and delivery of the mental health services.

For many years, mental health services have been provided in a variety of sites that may include hospitals, respite facilities, day-hospital settings, clinics, private offices, group homes, private homes, SRO's, shelters, and the streets. Providers of these services consist not only of traditionally trained professionals, nurses, nurse practitioners, occupational therapists, psychiatrists, psychologists, and social workers, etc., but also importantly now include consumers who may be employed as case managers or peer counselors (Curtis and Hodge, 1994). In fact, consumers *qua* providers may pose unique boundary considerations;

as Pollack discusses in the following chapter, there are particular dilemmas related to their dual roles, possible prior relationships, and special issues concerning confidentiality.

Moreover, the advent of managed care obfuscates the dividing line between providers and those who make administrative decisions, which are based on the availability of financial resources, about the provision of care (Backlar, 1996b). What a provider is able to do for a single client may depend on what she does for all her clients. To serve all her clients adequately, a provider needs to maintain a good reputation among those who make resource decisions. Indeed, her clients are not the only clients in the pool—she must share resources with her professional colleagues (Hollis, 1998). Needless to say, however, a client wants her best interest protected. The provider wants this too, but she also aspires to do the best for *all* her clients (Backlar, 1996a). Consequently, providers' special obligations to their individual consumer/patients as customarily viewed—e.g., the continuity of the relationship between the provider and the consumer/patient which has long been considered of foremost importance (due to the nature of mental health disorders)—does not always hold steady under a system of managed care (Wolf, 1994). In many circumstances today, providers may see their individual clients episodically, or providers may be used interchangeably. When providers and their clients are strangers to each other, boundary guidelines, concerned with maintaining a decorous distance between the parties, may no longer pertain. In truth, novel approaches that effectively establish an immediate and close trusting relationship are more likely to be considered.

Providers may assume that dilemmas in relationship boundaries are intensified in rural settings. As Brownlee (1996) has observed, boundary issues pose some unique problems for rural mental health providers when low density populations may foster mutually dependent relationships. In a small town, the client's aunt may be the clinic's receptionist, the provider's husband may be the client's son's school teacher, or the provider's sister may be the client's husband's boss. But, urban community mental health settings that offer a variety of community support programs may also have corresponding complications. The actual cases may be different but upon careful inspection relationship boundary dilemmas may end up being quite similar (Curtis & Hodge, 1994; Diamond & Wikler, 1985). For instance, in a metropolitan area, a client with a severe and persistent mental disorder may decompensate and refuse services, yet her treatment team (which may include a consumer peer counselor with whom she has had some previous acquaintanceship) may decide for therapeutic reasons to establish and maintain a close contact with her family, her landlord, her local grocery store, and her pharmacy.

We are likely to find that concerns relating to relationship boundaries abound equally in both rural and urban locations. Health care providers' disquiet about this subject is revealed in the thriving literature on relationship boundaries (Brownlee, 1996; Curtis & Hodge, 1994; Epstein, 1994; Goisman & Gutheil, 1992; Pam, 1994). The many-sided and divers nature of today's community mental health support services may serve to confuse the healthcare professional's traditionally held views about her role, about her relationships with her clients, and about her understanding of common boundary rules. The conventionally trained professional provider—whose education has underscored the scientific perspectives of knowledge, and whose clients are apt to present themselves voluntarily for treatment—may only be at ease if she limits her practice to a "professional context," e.g., within the confines her office (Alvarez, Batson, & Carr, 1976). However, a provider may not have the luxury of waiting for clients to visit the clinic. Depending upon her clients' needs and wants, she may be asked to take her clients out for coffee, to meet with her clients in their homes, and to help them with their grocery shopping. Indeed, a provider may be required to approach mentally ill persons who are lying on park benches, huddled under bridges, or sleeping in downtown storefronts. She may offer such persons food, clothing, blankets, and other such items, in an attempt to gain their trust—her long term goal being to secure them medical and psychiatric treatment, permanent housing, and other social services.

Of course, some judgments in regard to relationship boundaries in community support services may be quite straightforward.We all are likely to agree that the following boundary taboos should be maintained: a provider should never have sex or engage in intimate physical contact, should never physically, verbally, or emotionally torment, should never use drugs with, provide drugs to, or purchase drugs from, any person to whom she furnishes mental health and/or support services (Curtis & Hodge, 1994). Less dramatically, most relationship boundary dilemmas are likely to result from the prosaic, everyday, and frequent contacts that occur between a provider, her clients, and the sundry persons who may have some connection to a particular provider/client dyad. And more often than not, some of these challenges may be intensified when the provider finds herself pulled between her professional role and her personal self.

Most of us have rather distinct notions about the role that professionals play. For instance, if I mention words like nurse, doctor, social worker, or police officer, we may tend to conjure up popularized images like a man in a white coat with a stethoscope, or a woman, also in white, with a bed pan, and so forth. It appears natural that a provider may also have some generalized and universalized—though more sophisticated—ideas about her professional role. I suspect there may be some tension that exists between her conventional view of her professional role and her own particular and personal values. Probably, her view of her professional role is predicated on a professional code that endorses what Hollis (1988) terms an "agent-neutral ethics." This would be a golden rule, a general principle, an impersonal type of ethic: do as you would be done by; Kant's categorical imperative—"an action is right for a person if it is right for anyone so placed"; or, a utilitarian ideal—"we ought to promote the greatest good of the greatest number." While antithetically, a provider's own partial and personal values may stem from an individual type of ethic that counts each person to be morally distinctive (Hollis, 1988). More than likely, a provider would describe herself as a woman whose determinations and projects define her personal life; she treasures her liberty to make decisions; and she appreciates that disparate people hold dissimilar values in similar situations. Her distinctive and unique experiences and her moral values may underlie her professional decisions, e.g., to violate a client's confidentiality because her judgement tells her that in this instance it is correct and right to break a certain "rule," or, in other circumstances, to befriend a client whose needs prompt her to disregard conventional boundaries. Upon occasion, she may feel be torn between the impersonal standpoint that demands a universal impartiality and her particular, personal standpoint, which can be—but need not be—corrupted by self interest or custom (Nagel, 1991).

We should appreciate that the care-giving provider has a demanding and often arduous responsibility. Whatever the circumstance, she attempts to be both involved and objective. She is required to recognize who is a vulnerable person; she must identify the essential needs of that person; in order to address those needs she must endeavor to gain that person's trust; and she must never take advantage of that trust. However, there are no codes of ethics or training that can capture her everyday moral decision-making and reduce it to a model strategy. Learning *rules* about roles and relationships may be comparable to our school-room instruction of reading, writing and arithmetic. The three Rs are tools that aid our efforts to communicate with others, but they do not provide us with a technique for resolving our human dilemmas. How to fittingly use rules and skills requires personal judgement. In the absence of such an ability there is not much that can guarantee against wrong employment. We are capable of understanding instructions and learning rules, but "judgement is a peculiar talent that can be practiced only, and cannot be taught" (Kant, [1781] 1965).

## REFERENCES

Alvarez, R., Batson, R. M., & Carr, A. K. (1976). *Racism, elitism, professionalism: Barriers to community mental health.* New York: Jason Aronson, Inc.

Appelbaum, P. S., & Jorgenson, L. (1991). Psychotherapist-patient sexual contact after termination of treatment: An analysis and a proposal. *American Journal of Psychiatry, 148*(11), 1466–1473.

Appelbaum, P. S. (1990). Statutes regulating patient-therapist sex. *Hospital and Community Psychiatry, 41*(1), 15–16.

Backlar, P. (1996a). The three Rs: Roles, relationships, and rules. *Community Mental Health Journal, 32*(5), 505–509.

Backlar, P. (1996b). Managed mental health care: Conflicts of interest in the provider/client relationship. *Community Mental Health Journal, 32,* (2) 101–106.

Brownlee, K. (1996). The ethics of non-sexual dual relationships: A dilemma for the rural mental health professional. *Community Mental Health Care Journal, 32*(5), 497–503.

Curtis, L. C., & Hodge, M. (1994). Old standards, new dilemmas: Ethics and boundaries in community support services. In L. Spaniol, M. A. Brown, L. Blankertz et al. (Eds.). *An Introduction to Psychiatric Rehabilitation.* The International Association of Psychosocial Rehabilitation (IAPSRS), pp. 339–354.

Diamond, R. J., & Wikler, D. I. (1985). Ethical problems in community treatment of chronically mentally ill. In L. I. Stein, & M. A. Test (Eds.). *The Training in Community Living Model: A Decade of Experience. New Directions for Mental Health Services, 26,* 85–93. San Francisco: Jossey-Bass.

Epstein, R. S. (1994). *Keeping boundaries: Maintaining Safety and Integrity in the Psychotherapeutic Process.* Washington, DC: American Psychiatric Press.

Gabbard, G. O., (2000). Boundary violations. In S. Bloch, P. Chodoff, & S. A. Green (Eds.). *Psychiatric Ethics,* 3rd ed. Oxford University Press 2000.

Gabbard, G. O., & Nadelson, C. (1995). Professional boundaries in the physician-patient relationship. *Journal of the American Medical Association, 273,* 1445–1449.

Goisman, R. M., & Gutheil, T. G. (1992). Risk management in the practice of behavior therapy: Boundaries and behavior. *American Journal of Psychotherapy, XLVI*(4), 532–543.

Gutheil, T. G., & Gabbard, G. O. (1992). Obstacles to the dynamic understanding of therapist-patient sexual relations. *American Journal of Psychotherapy, XLVI*(4), 515–525.

Gutheil, T. G., & Gabbard, G. O. (1993). The concept of boundaries in clinical practice: theoretical and risk-management dimensions. *American Journal of Psychiatry, 150,* 188–196.

Gutheil, T. G., & Gabbard, G. O. (1998). Misuses and misunderstandings of boundary theory in clinical and regulatory settings. *American Journal of Psychiatry, 155,* 409–41.

Hellman, S. (1995). The patient and the public good. *Nature Medicine, 1*(5), 400–402.

Hollis, M. (1988). A death of one's own. In J. M. Bell, & S. Mendus (Eds.). *Philosophy and Medical Welfare*. Cambridge: Cambridge University Press.

Kant, I. ([1781] 1965). Critique of Pure Reason. In N. K. Smith (Trans.), *Immanuel Kant's Critique of Pure Reason*. New York: St. Martin's Press.

Nagel, T. (1991). *Equality and Partiality*. New York: Oxford University Press.

Nussbaum, M. C. (1986). *The Fragility of Goodness: Luck and Ethics in Greek Tragedy and Philosophy*. Cambridge: Cambridge University Press.

Pam, A. (1994). Limit setting: Theory, techniques, and risks. *American Journal of Psychotherapy*, *48*, 432–440.

Wolf, S. M. (1994). Health care reform and the future of physician ethics. *Hastings Center Report*, *24*, 28–41.

CHAPTER 4

# RESPONDING TO BOUNDARY CONFLICTS IN COMMUNITY SETTINGS

DAVID POLLACK

## INTRODUCTION

As Backlar has stated in the previous chapter, additional meaning and complexity in relation to boundary issues have emerged with the increased diversity of circumstances, relationships, and locations in community programs and the transformation of care delivery systems. In community mental health settings, we have frequently been so overwhelmed with service demands and obligations to meet shifting regulatory requirements that we have not consistently attended to some of the ethical dilemmas that have developed. This is a mistake that we must avoid.

In a previous position as the medical director of a large urban community mental health agency, I was constantly confronted with a wide range of ethical dilemmas, either brought to my attention by other staff who became conflicted over how to deal with certain situations, through the complaints or reports of critical incidents, or through my own clinical activities. In addition to the usual boundary dilemmas, such as confidentiality concerns and the relationship limits between clinician and patient, that are seen in private practice settings (Appelbaum,

1982; Stone, 1983), a number of factors have contributed to a new range of ethical concerns in community settings (Christensen, 1997).

Some of these are related to the expansion of the workforce in community settings to include staff and volunteers who are not trained in the mental health professions and, therefore, have no formal exposure to professional codes of ethics (Schuster et al., 1994; Redick et al., 1996; Dracy & Yutrzenka, 1997; Williams & Swartz, 1998). The growing and positive movement to include consumers and families as active participants in their own and other clients' care unfortunately has added significantly to the types of boundary dilemmas that must be addressed (Paulson, 1991; Dixon, Krauss, & Lehman, 1994; Fisher, 1994; Haiman, 1995; Manning & Suire, 1996; Mowbray et al., 1996). The diversity of locations in which treatment services are provided, especially residential and milieu-based treatment programs (Bachrach, 1994), not to mention the in vivo treatment experiences that are provided by outreach clinicians and mobile crisis response teams, blur the boundaries of the service frame and necessarily compromise the privacy of such interactions.

How do we recognize and respond to such difficult situations, especially with staff who may be less well trained in ethical practice and operating under demands to provide a wide range of high quality services to clients with ever increasing pressure to contain costs and to provide sufficient documentation for accountability purposes?

## ONE CMHC'S RESPONSE

Over a period of several years, we organized an effort to increase staff awareness of these ethical issues, with particular emphasis on boundary concerns. We created an ethics committee, with representation of clinical and administrative staff from throughout the agency, which has met monthly to identify ethical concerns, develop policy and training recommendations, and monitor and discuss specific dilemmas as they arise. Early on in the course of the ethics committee, we determined that some baseline training needed to be provided to all staff, with specific emphasis on promoting the best interests of the client and the importance of respecting the relationship boundaries that are so critical to the well-being of the client (Wells, Hoff, & Benson, 1984). We also decided that it was essential to provide a set of ethics guidelines for all staff and took several months to draft them. These guidelines were ultimately included in the orientation packet for all new staff, who would review them with their supervisors. The guidelines were intended to alert staff to potential ethical conflicts, to promote the notion that it is perfectly acceptable to ask for help in resolving them, and to provide suggestions for how to get such help. Our experience resembled Backlar's observations: we created very few absolute rules in the guidelines, because there are so many complicating situations with many possible responses, depending on the circumstances. An example of such guidelines is included as an appendix to this chapter.

Throughout the time that the ethics committee existed, some of the most valuable experiences were the case discussions or ethics rounds that occurred (Appelbaum & Reiser 1981; Reiser et al., 1987). The open atmosphere and support for staff to raise and discuss such difficult situations led to us recognizing and responding to a number of previously unanticipated problems. One of the more difficult dilemmas had to do with the boundary conflicts associated with consumers who became employees within the agency. As more and more clients went through our peer counseling training program, several of them were recruited to work in various capacities within the agency, as peer advocates, outreach workers, day treatment staff, and case management aids. The problems associated with being in dual roles, having prior and continuing relationships with clients who were now being served by the consumer-employees and continuing to receive services from staff with whom they may now have collegial and labor relationships, led to lengthy and complicated discussions. Eventually we resolved to recommend that consumer employees consider receiving clinical services from a different agency or provider than the one for which they worked. We also created a set of questions for prospective consumer-employees to consider prior to seeking or accepting such employment. These questions reflected our desire for the consumer to be aware of potentially restrictive personnel policies, to consider the impact on current and previous relationships, and to be alert to the risks of increased symptoms secondary to the pressures brought on by the work situation. We were clearly committed to supporting consumers getting appropriate, rehabilitative, and empowering employment opportunities. At the same time, we wanted to guard against a myriad of problems, about which we could only speculate.

As other ethical concerns arose, the same pattern of open discussion, with consideration of the various ethical principles involved, and productive group-based problem solving, led to other policy recommendations, client and clinician guidelines, and increased awareness to future dilemmas.

## CASE SCENARIOS

The remainder of this chapter is an attempt to portray the range and complexity of the ethical dilemmas encountered in community practice. In addition to understanding that one cannot devise many absolute rules for ethical behavior, it is also not possible to provide a detailed description of problems and potential solutions. We found that "learning by doing", identifying real life situations and dilemmas and then discussing them in a collaborative and supportive manner was extremely helpful.

The following case scenarios and derivative questions are presented in such a way as to provoke discussion and to demonstrate that there are no unambiguous solutions to these dilemmas. Discussion of the specifics of such scenarios can lead to effective policies and improved skills and attitudes in administrative and clinical staff so that they will be better equipped to respond to future ethical conflicts.

### CASE 1

J. S., a client with severe and persistent mental illness (SPMI) at a CMHC, has been stable for several years and has gone through peer counseling training in the psychosocial rehabilitation program there. Now J. S. has graduated from that program and has applied to be a case management aid for one of the teams in the same agency. He clearly meets the qualifications for the position and, in fact, was the best candidate for the position, surpassing several other applicants who do not have evidence of psychiatric illness. In discussing the possibility of hiring J. S., the program manager suggests that he may encounter resistance from some staff, who know that he lives with two other persons, both of whom are clients at the same agency and who receive services from the team for which J. S. would be working. J. S. also has been told that he would probably be better off if he transferred his own care to another agency on the other side of town, but he doesn't like the idea of leaving the primary clinician with whom he has worked for so many years and who has supported him through several major crises. J. S. decides to accept the position and transfers his care to the other agency. After starting to work, he begins to experience some increased symptoms of persecutory ideation. He attributes this to the facts that several of the clinicians stopped coming to union meetings after he had become a union employee and that his supervisor has told him that he cannot have a key to the building nor access to the clinical records area without permission.

Why should we be concerned with consumer-employees receiving services from the same organization with which they work? Should we require consumer-employees to adhere to the same personnel policies as other employees? If we do, then shouldn't such employees be given the same privileges (e.g., keys) as the rest of us?

## Case 2

G. B. is the recreation therapist in a psychosocial rehabilitation program in a large CMHC. She received formal training in recreational therapy, has worked in social service agencies for many years, and has a reputation for being very personable and friendly with clients. Her strong beliefs in counter-culture philosophical issues such as herbal medicine, massage, and vegetarianism are well known among her peers, and they have not been a problem for her in her work. In fact, some clients have formed a group, which she leads, within the program to learn more about natural foods and herbal remedies. However, when she is on a field trip to the coast with two other staff and six clients, she encourages the group to go to a local hot springs and she disrobes in front of the group. She is surprised when her coworkers insist that they put their clothes back on and leave the hot springs. She insists that the experience was intended to allow the clients to express their pent up feelings of being caged and inhibited by their illness and the way others treat them as children.

Some non-traditional and less conventional clinical approaches cross traditional boundaries. How can one tell if such boundaries have been inappropriately breached and who should intervene? Is GB's explanation of her attempt to liberate clients justified or merely a rationalization of her indulgence of countertransference feelings for one or more of the clients? What if the clients viewed the disrobing experience as "normal" and acceptable for the hot springs context and became insulted at the paternalistic or infantilizing attitude of the other staff?

## Case 3

L. M., a retired accountant, is the parent of a schizophrenic young man. He is very concerned about his son and how many times he has been hospitalized. His son is currently doing fairly well in a case management program in which his money is managed and he is living in a structured group home. L. M. becomes a volunteer at the agency in which his son is receiving services, offering his accounting skills wherever they can be used. His volunteer program supervisor, who is excited to have the services of a skilled volunteer, doesn't know that L. M.'s son is a client in the program. She assigns L. M. to work with the money management program reviewing and reconciling their books. In the process of working in this area, L. M. discovers how much money his son gets each month and how it is spent. L. M. has had much conflict with his son in the past, mainly because his son has refused to cooperate with L. M.'s attempts to get him into treatment, complaining that L. M. was "trying to control my life". He calls his son to suggest that he not spend as much on cigarettes.

What can we do to adequately screen volunteers (or employees) for potentially compromising connections to other staff or clients? It is difficult to recruit enough volunteers and to find those who are qualified to do difficult or highly skilled tasks. This is due to the overall shortage of volunteers and the stigma and fear associated with working with persons with mental illness. Do we risk losing potentially valuable volunteer resources with the imposition of such screening? Should L. M.'s son or his primary clinician have been informed of the volunteer assignment and given the opportunity to veto the decision?

## Case 4

The mobile outreach team of an urban CMHC spends much of its time identifying and developing relationships with people who appear to be homeless and mentally ill. When they make contact with a prospective client, they try to encourage him or her to accept food, shelter, and to receive certain mental health services, such as evaluation and medications. Only when they perceive that to do so would significantly jeopardize their safety or would scare off a potentially receptive client, the team will not initially identify themselves as mental health workers. One client they have been following for many months tends to be very isolated and suspicious. She eats out of dumpsters, but occasionally goes to one of several downtown churches for food and shelter in their basement for a few nights at a time. In fact she regularly rotates among these different churches, eliciting concern from the staff in each church until her disorganized and verbally assaultive behavior get her to the point that she leaves the church, thinking that they are plotting against her. The outreach team has not been able to succeed in getting her into services, because she manages to survive on these rotating handouts from the churches. The outreach team clinician meets with the staff from all of the churches and convinces them to not help this woman, but to urge her to come to the outreach team for help.

Should prospective clients be informed that the outreach workers are mental health providers? How much dangerousness or deterioration should a person exhibit before such outreach interventions are justified? What are the problems associated with engineering a solution in which care or support is denied or refused in order to influence a person (who is not a willing or acknowledged "client") to get into more appropriate or effective care? Do we risk crossing civil libertarian or legal boundaries by developing such arrangements?

### CASE 5

R. F., a client who has been in the hospital many times in recent years, usually because he stops taking his medications, is now living in a structured residential program. He likes his room and the free meals that are part of the program. He agreed at the time of discharge from his last hospitalization to receive money management and to have his medications stored in the office of the residential program. Ten weeks after discharge from the hospital, he begins to show signs of decompensation. It is discovered that he has been "cheeking" his medications for the past 2–3 weeks. His primary clinician arranges with the money manager to not give R. F. his thrice weekly allowance of money for cigarettes and incidental spending money on any day that R. F. doesn't brings in a note from the residential program saying that he took his medications without "cheeking" them.

When is it acceptable for us to use leverage to get voluntary clients to comply with treatment and how do assure that it is not being done for expediency or other inappropriate reasons associated with the need to control the client? What conflicts do we create by handling the client's money? Would the situation be altered if R. F.'s money management agreement clearly stipulated the requirement for him to take his medications without cheeking them before receiving his daily money allowance?

### CASE 6

A large CMHC has recently entered into a managed care contract with the state in which services are paid on a capitated basis, i.e., they receive a certain fixed amount of money per enrolled subscriber in a certain health plan. They must use this budget to provide all services to the clients who come to their program and cannot bill for any of the services they provide. The psychosocial rehabilitation program at the agency has sustained significant cuts because of the reduced funding in this new budget arrangement. The program manager in this area decides to recruit 10–15 clients to work as "voluntary" case management aids. She determines that these clients can be effective companions for some very disturbed clients and can help provide interpersonal skills training. When someone raises the concern that this might be stressful work for these client-volunteers, she counters that they will meet once or twice each week with a case manager, who will be able to support them and identify any of them who are not up to working with this group of clients.

When is the use of consumers as volunteers or employees justified and when is it simply exploitation? What if such arrangements are constructive and supportive of consumer-employees and developed with their full approval and cooperation, but are at least partially motivated by limited resources or threats of funding cutbacks? How can we ethically reconcile and balance our commitment to consumer employment with our need to cut costs?

## CASE 7

A CMHC has entered into a managed care contract similar to the one described above. However, as one means of containing costs, this program has arranged to share some of any budget surplus with the clinical staff. This means that if, at the end of each quarter, the agency's costs are less than the budgeted amount, some of the savings will be returned to the staff in the form of bonuses. Some staff are uncomfortable with this arrangement, claiming that they are being encouraged to provide less care in order to save money and that some clients will suffer. The program manager counters that they must not cut corners too much because, if the clients don't do well and end up being hospitalized, those costs are borne by the agency, thus making it less likely that there would be budget surpluses.

Can financial incentives ever be constructed which do not promote increased risk of ethical compromises and, if not, how can such risks be minimized or monitored? What about the ethical problems created when we provide excessive, unnecessary, or redundant services? Fee-for-service arrangements create the incentive to "do more so that we can bill for more." Is such "excess" service justified if it provides a necessary subsidy for other essential, non-revenue producing services (such as mobile outreach or intensive case management)?

## CONCLUSION

The foregoing discussion and examples represent one CMHC's experience in dealing with these very troubling and complicated problems of changing and multiple roles, blurry relationship boundaries, and the absence of clear and unambiguous rules to guide clinician behavior (Wasow, 1999). By promoting an open and accepting atmosphere for collegial inquiry and discussion, encouraging staff to seek support in resolving their ethical dilemmas, and emphasizing the primacy of the best interest of the client, we found that we had fewer disturbing incidents and swifter, more effective responses to ethical conflicts. This "learning by doing" approach was well accepted and effective. It is important to avoid the pitfall of excessive deliberation in the face of ethical conflict. Program staff and supervisors should act consciously, but with a readiness to alter course, acknowledge mistakes, and share results.

As helpful as this attention to ethical issues has been, a troubling question remains: can the expense of such efforts be encouraged, tolerated, or even allowed in the increasingly cost-conscious environment of managed care (Lazarus & Pollack, 1997)? We certainly think that it should and that this administrative expense should be factored into any program's budget as an appropriate component of quality assurance.

## APPENDIX

### Sample Mental Health Organization Ethics Guidelines

1. *Introduction*
These guidelines set forth the basic ethical principles that apply to all staff of the agency. They are intended to provide a framework for making decisions about ethical conduct and a reference when questions arise. Staff members should also be familiar with the ethical codes of their professions (if applicable) and with the various other policies and procedures of the agency that pertain to their work.

2. *Respect for Human Dignity*

a. Do not discriminate in providing services or access to programs on the basis of race, gender, age, national origin or culture, religion, sexual orientation, disability, or socio-economic status.

b. Show respect for the dignity and worth of all individuals.

3. *Professional Behavior*

a. Always act in the best interests of the client.

b. Know your role as an employee of the agency and the limits of your expertise. Act within those limits and within the confines of your training and your position in the agency. For example, some staff have positions which do not include discussing personal or intrapsychic information with clients. If in doubt, consult with your supervisor. It is important to know when to be a good listener and when to tell a client that a discussion is not one you should be having.

c. Avoid harmful multiple relationships or any behavior that is not consistent with the client's treatment. Examples:

- Do not drink alcohol or take drugs with clients. Any discussion of alcohol or street drugs should be consistent with the treatment plan. Providing alcohol or street drugs to clients will result in immediate termination.
- Do not date clients. Any staff member who engages in sexual relations with a client of MHSW will be dismissed immediately. Personal relationships with clients outside of work hours are discouraged.
- Do not buy from or sell to clients. Many of our clients are without funds and feel the need to sell personal items at low cost. Staff must not exploit these situations. For the same reasons, staff generally should not make personal loans to clients. (Agency petty cash may be used for loans in an emergency.)
- Any gifts to or from clients should be discussed with your supervisor.

4. *Confidentiality*

a. All client records, including identity, must be protected with extreme care.

b. Do not discuss any information regarding a client with anyone outside of the clinic without authorization.

5. *Clinical Ethical Principles*

The following areas mainly pertain to the primary clinician, but everyone should understand what these terms mean.

a. Informed consent. Clients have the right to be informed of and to understand the nature of their treatment and the risks and consequences associated with that treatment.

b. Client abandonment. Clients should not be terminated from services without providing appropriate referral and/or support.

c. Duty to warn. In certain circumstances where there is a specific expressed intent to harm another person, there may be an obligation to inform the affected parties of the threat.

d. Duty to report. There is an obligation to report suspected abuse of certain vulnerable populations, including children and the elderly.

6. *Resolution of Ethical Issues*

a. There should be ongoing discussion of ethical questions. When in doubt, the matter should be discussed with your supervisor, the client's primary clinician, or another qualified staff member with expertise in ethical issues.

b. If you have concerns about the ethical behavior of others within or outside of the agency, you should normally raise the issue with your supervisor or another qualified staff member. It may be appropriate to discuss the issue with the person involved or report the matter to an outside party. If in doubt, consult with your supervisor or an Ethics Committee representative.

## REFERENCES

Appelbaum, P. S. (1982). Confidentiality in psychiatric treatment. In L. Grinspoon (Ed.) *Psychiatry 1982: The American psychiatric association annual review*. Washington, D.C.: American Psychiatric Press.

Bachrach, L. (1994). Residential planning: Concepts and themes (The chronic patient). *Hospital and Community Psychiatry, 45,* 202–203.

Christensen, R. C. (1997). Ethical issues in community mental health: Cases and conflicts. *Community Mental Health Journal, 33,* 5–11.

Dixon, L., Krauss, N., & Lehman, A. (1994). Consumers as service providers: The promise and challenge. *Community Mental Health Journal, 30,* 615–625.

Dracy, D. L., & Yutrzenka, B. A. (1997). Responses of direct-care paraprofessional mental health staff to hypothetical ethics violations. *Psychiatric Services, 48,* 1160–1163.

Fisher, D. B. (1994). Health care reform based on an empowerment model of recovery by people with psychiatric disabilities. *Hospital and Community Psychiatry, 45,* 913–915.

Haiman, S. (1995). Dilemmas in professional collaboration with consumers. *Psychiatric Services, 46,* 443–445.

Lazarus, J., & Pollack, D. A. (1997). Ethical aspects of public sector managed care. In K. Minkoff, & D. A. Pollack (Eds.) *Managed mental health care in the public sector: A survival manual*. Newark, NJ: Harwood Academic Publishers.

Manning, S. S., & Suire, B. (1996). Consumers as employees in mental health: Bridges and roadblocks. *Psychiatric Services, 47*, 939–943.

Mowbray, C. T., Moxley, D. P., Thrasher, S., Bybee, D., McCrohan, N., Harris, S., & Clover, G. (1996). Consumers as community support providers: Issues created by role innovation. *Community Mental Health Journal, 32*, 47–67.

Paulson, R. (1991). Professional training for consumers and family members: One road to empowerment. *Psychosocial Rehabilitation, 14*, 69–80.

Redick, R. W., Witkin, M. A., Atay, J. E., & Manderscheid, R. (1995). *Human resources of mental health organizations: United States, selected years 1972–1990*. Rockville, MD: Center for Mental Health Services.

Reiser, S. J., Bursztajn, H. J., Appelbaum, P. S., & Gutheil, T. G. (1987). *Divided staffs, divided selves: A case approach to mental health ethics*. Cambridge, MA: Cambridge University Press.

Schuster, J. M., Kern, E. E., Kane, V., & Nettleman, L. (1994). Changing roles of mental health clinicians in multidisciplinary teams. *Hospital and Community Psychiatry, 45*, 1187–1189.

Stone, A. A. (1983). Sexual misconduct by psychiatrists: The ethical and clinical dilemma of confidentiality. *American Journal of Psychiatry, 140*, 195–197.

Wasow, M. (1999). Sure we believe in ethics and values: But whose? *Community Mental Health Journal, 35*, 489–492.

Wells, K. B., Hoff, P. A., & Benson, M. C. (1984). A medical ethics tutorial program. *Journal of Medical Education, 59*, 433–435.

Williams, J., & Swartz, M. (1998). Treatment boundaries in the case management relationship: A clinical case and discussion. *Community Mental Health Journal, 34*, 299–311.

Part III

# VIOLENCE AND MENTAL DISORDERS

Chapter 5

# AT HOME WITH THREATS AND VIOLENCE

Patricia Backlar

*Few are interested in either heightening the stigmatization of the mentally ill or impeding progress of the mentally ill in the community. Yet this progress is bound to be critically slowed without a realistic look at dangerousness.*

—J.D. Bloom (1989, p. 253)

In the early 1980s, a newspaper article that described the rape and murder of a woman by her son who had schizophrenia, caught my attention (Tax lawyer's son committed, 1983). A few years previously, my own son had been diagnosed with schizophrenia. Consequently, I read the article quite carefully. The young man lived at home with his parents. The father, on returning from work at the end of the day, noticed his wife's broken glasses near the front door prior to discovering her body in their bedroom. The personal detail about the glasses evoked for me considerations about the tragedy that were not reported. I wondered if the mother and father had been afraid of their son. And if so, how they balanced that fear with their love for him, their worry about his illness, their uneasiness about his dependency upon them, and their concern for his future. I worried about the young man himself: did he now—or would he later—experience pain, torment, and loss? what lay ahead for him? and would his father be able to continue to love him and protect him? I mulled over the calamitous losses this family had endured: a mother and wife defiled and dead, a husband and father bereaved, and a son dispossessed. In the classical tragedies, of which this family's catastrophe is so reminiscent, a *deus ex machina*—a "god from a machine"—was an unexpected event, a gift from the gods, introduced to resolve a formidable situation. Could this family's misfortune, I questioned, have been prevented? Was this case an instance of bad luck or were most people with severe mental illnesses predisposed to violent acts? Would my own son be capable of such violent behavior? How could I protect him? How could I protect myself?

In recent years, our society has shown an escalating interest in and concern about violence in the home. The major focus has been on child and spousal abuse with some passing nod to elder mistreatment. However, little or no attention has been paid to the despair and fear of families with mentally ill relatives, who may live lives fraught with intermittent tumult, the threat of violence, and actual violence (Backlar, 1994; Isaac & Armatt, 1990). Paradoxically many of these families are dedicated to protecting their relatives from harm, while—at the same time—they may be in jeopardy themselves. Typically, their anxiety for their son, daughter, father, or mother, outweighs their anxiety for themselves.

That there appears to be some kind of relationship between serious mental illness and violent behavior has been no secret. Historically, this belief has been observed to be both culturally omnipresent and unchanging (Monahan, 1992a,b; Murphy, 1976) and has received plenty of attention in literature, the popular press, and (in the past century) in film and television. Many families—who have relatives with mental illness—are only too well aware of the connection (Backlar, 1994; Steinwachs, Kasper, & Skinner, 1992; Torrey, 1994). Yet, families may be reluctant to admit that there is an association between mental disorder and violence. Indeed, consumer advocates, mental health professionals, and social scientists, also may downplay or deny—for a variety of differing reasons—that there is any such relationship. I suspect that all these parties (despite their discrete divergent views about, and special interests in, this issue) share a desire not to perpetuate society's oversimplified negative characterizations about persons who suffer from serious mental disorders.

In the past and today, people who have certain kinds of diseases such as leprosy, TB, AIDS, mental disorders, etc., often are treated as outcasts by the societies in which they live (Backlar, 1994). Such stigmatization can make everyday life both hard and sorrowful for mentally ill persons, and consequently for their families: finding adequate places to live, securing jobs, and gaining access to medical care can be onerous or even impossible. To make matters even more difficult, it has become popular to recklessly bandy about the phrase "dysfunctional family." Families, so characterized, are perceived as self-sustained disordered systems that live in a wretchedness of their own making. A family's impaired functioning is likely to be seen as causing the anti-social actions by an individual member of the family. And circularly, each individual member's actions are seen to reflect the impaired functioning of the family entity itself (Lefley, 1989). Because it appears to be a common human wish to be well thought of by others, many of us strive to establish a "good name." And, should we lose such a prized good as our reputation, we may be quick to feel a sense of shame (Rawls, 1971). Families are loath to bring dishonor to themselves and their mentally ill relatives. Understandably, families may refrain from calling attention to their relatives' threatening or violent behaviors. Such reluctance, however, may in the long run not only seriously disadvantage their ill kin, but also may cause severe harm to themselves.

In a notable presentation and paper, John Monahan (1992b), reveals that although previously he and many social science researchers (along with advocates for mentally ill persons) claimed that there was no fundamental relationship between mental disorder and violent behavior, he now believes that there may be a relationship "that cannot be fobbed off as chance or explained away by other factors that may cause them both." The position that Monahan and others formerly had taken (in which they denied a special relationship between violent behaviors and the severe mental illnesses) may have been driven, in part, by their concern that such a presumption would stimulate both disparaging public and private attitudes toward, and discriminatory legal and health policies for this particular population. Moreover, there was a class of evidence that may have reinforced such a cautious position. Prior to 1965 most people with severe mental illness resided in mental hospitals (Grob, 1991). Therefore, as cited in Torrey (1994), studies up until 1945 all confirmed that "mentally ill persons had a lower arrest rate than the general population" (Brown, 1985). However, even though arrest rates are easily obtained and somewhat germane, they are an unsophisticated measure, particularly at a time when most mentally ill people were locked up (Shah, 1990). Furthermore, it was common knowledge that before the advent of anti-psychotic medications rampant violent behaviors were common on the wards of mental hospitals (Isaac & Armat, 1990).

I suspect that the advent, in the 1960s, of the rights-oriented movement with its concerns about the welfare and rights of prisoners, mental patients, women, and human subjects in research protocols (Rothman, 1987; Rothman, 1991), powerfully influenced Monahan and colleagues' attitudes. In addition, concerns about civil-rights animated efforts to move patients out of the state mental hospitals, thus fueling a plethora of mental health law and litigation. The 1967 California law, the Lanterman-Petris-Short (LPS) Act (California Welfare and Institutional Code 5000–5464), provided a model for moving involuntary commitment decisions away from a position based on the legal doctrine of *parens patriae* (which authorizes the state to grant protection for those who cannot care for themselves) toward an emphasis on "danger to others" and procedural rights (Scull, 1989). The LPS Act was devised in order to protect mentally ill persons' civil rights by putting a stop to indefinite commitments. Indeed, as Grob (1991) notes, it was a law that pleased both liberals and conservatives alike: while appearing to promote the welfare of mentally ill people, it also was a preliminary step toward the demise of state mental hospitals, which were operated at a great cost to the state.

With the passage of the LPS Act, the concept of "danger to others" was forced onto center stage, connecting the notion of violent behavior to persons with mental disorders. For people—like Monahan—who well-perceived the negative and destructive consequences that stigmatization can have upon the lives of persons with severe mental disorders, the LPS Act, although ostensibly attending to this population's wellbeing, had undesirable side-effects: First, by attaching the idea of "danger to others" (and making it a pivotal issue) to civil commitment proceedings for people with mental disorders, anecdotal myths and prejudicial folk beliefs that stereotype this population were likely to be reinforced; and second, these folk opinions—now strengthened—could deleteriously influence policies and laws designed to keep this population "in line," and also the ways in which mental health care would be administrated. Such concerns may have colored many researchers' judgements, and determined their commitment to deny a connection between mental disorder and violent behavior.

Ironically, the 1960s civil-rights activities, which hastened the deinstitutionalization of state mental hospital patients, actually created the social circumstances that made possible broad epidemiological research studies. No longer confined for extended periods of time in mental hospitals, most severely mentally ill persons (who were not in jail or in prison) were living in the community—with their families, in group homes, in apartments, in single rooms, or on the streets. Thus facilitating the kinds of careful studies by Swanson and associates (1990), and Link and associates (1992) that provided the facts which altered Monahan and other researchers' previous opinions. The Swanson (1990) and Link (1992) studies revealed that mentally ill persons

> actively experiencing serious psychotic symptoms—are involved in violent behavior at rates several times those of nondisordered members of the general population, and that this difference persists even when a wide array of demographic and social factors are taken into consideration. Because the studies were conducted using representative samples of the open community, selection biases are not a plausible alternative for their findings (Monahan, 1992b).

Torrey (1994), also citing the Link and associates 1992 study, echoes Monahan's emphasis that a sub-group of persons with some history of mental disorder have significantly more incidents of violent behavior than community residents who had no such histories, that demographic and socioeconomic variables had little bearing on the differences between the two groups, and that it was the level of psychotic symptoms that accounted for the differences. "The sicker the patients," notes Torrey, "the more likely they were to have exhibited violent behavior." Other factors that add to the risk of involvement in a violent act for individuals in this population may include: having a dual diagnosis—i.e., having both a mental disorder and a substance abuse disorder (Mulvey, 1994; Swanson, Holzer, Gnaju, & Jono, 1990); and, experiencing command hallucinations—e.g., recognizing the hallucinated voice as a particular person or being, and having delusions that correlate with the hallucinations (Junginger, 1995). Furthermore, according to Junginger (1995), "dangerousness is not only a function of various characteristics of a patient and his or her illness, but also of the patient's psychotic experience that are based in the patient's environment." Consequently, a patient who appears to be stable in the hospital, may nevertheless experience critical command hallucinations in the post-hospital environment that could trigger violent behavior.

Generally, the number of violent acts by both male and female persons with mental disorders appear to be fairly equal; the differences being related more to the opportunity availed by the location where the person may be living (Newhill, Mulvey, & Lidz, 1995). For example, persons living alone are more likely to be involved in non-family assaults, whereas persons living at home may be violent with family members. And the literature (Binder & McNeil, 1986; Cook, 1988; Estroff, Zimmer, Lachicotte, & Benoit, 1994; Lefley, 1987; Straznickas, McNeil, & Binder, 1993) suggests that because some mothers' shoulder a substantial responsibility for the care of their mentally ill relatives, they appear to be at a heightened risk for becoming the target of their relatives' violent assaults.

The information gleaned from this recent research reveals a particular relationship between violent behavior and mental disorder. The previous belief that there was no consequential connection between mental illness and violent behavior in the community appears to be erroneous (Mulvey, 1994). Most significantly, however, as Monahan (1992b) underscores, no meaningful distinction was found in the rates of recent violent behavior between non-mentally ill (i.e., never-treated) community residents and persons with mental illness when "current psychotic symptoms were controlled." Monahan's account suggests that if psychotic symptoms can be managed—kept at bay—violent behavior in this population will occur at no greater rates than in the population at large.

Most of us are likely to agree that methods, which manage an individual's psychotic symptoms (prior to the point of becoming dangerous) in the community, are preferable to systems that promote incarceration in hospitals or jails. Yet, tragically as things stand today, jails and prisons have become the "nation's new mental hospitals" (Butterfield, 1999). However, even if predictive testimony about future violence turns out to have sufficient scientific support and becomes acceptable to psychiatry, psychology and the courts (Grisso & Appelbaum, 1992)—thus, in effect, validating methods to manage psychosis that involve the use of antipsychotic medications—the right to refuse medications will be evoked. Should that be the case, concerns about coercive actions that threaten consumers' freedom are apt to remain prominently featured in the discussion.

Commonly, we characterize coercion (Hiday, Swartz, Swanson, Borum, & Wagner, see chapter 8) as a type of influence that takes place when a person (usually in a position of authority) deliberately employs a plausible and serious threat of power or injury to control another person (Beauchamp & Childress, 2001). In a typical dispute, paternalists claim that coercive management may be tolerated if it prevents people from harming themselves. On the other side, civil libertarians contend that people's freedom may not be curtailed unless they have committed a crime or there is imminent danger of their harming others: the civil libertarian contends that it is acceptable to obstruct a person's liberty in order to protect others from harm, but not acceptable to protect someone from harming herself. According to Macklin (1993), a rigorous libertarian would go even further and admit only to the curtailment of a person's freedom when he has *already* committed a criminal act. Nevertheless, John Stuart Mill ([1859] 1977), who famously opposed paternalism, never envisioned that it could be considered paternalistic to interfere with the liberty of people who lacked the capacity to comprehend the consequences of their actions. In Mill's view, authentic human autonomy is based upon the faculty for rational agency (Backlar, 1995b). Indeed, some commentators argue that paternalistic decisions made for a person who is severely injured or is seriously mentally ill may be justified if "with the development of rational powers the individual in question will accept our decision on their behalf and agree with us that we did the best thing for him" (Rawls, 1971, p. 249). Such "retroactive consent" has been referred to as a "thank you theory" (Culver & Gert, 1982); Beauchamp and Childress (2001, p. 68) contend that this assumption should not be considered a replacement for informed—and voluntary—consent: "It is merely an anticipated outcome that reassures health professionals that they are acting in the patient's 'best interests'."

There are a number of approaches that theoretically may provide ways by which a consumer's psychotic symptoms could be controlled in the community. We are quite familiar with two of these approaches, the Program for Assertive Community Treatment (PACT) and involuntary Outpatient Commitment (OPC) (Hiday et al., see chapter 8). Yet, because informed consent processes are rarely used in either of these approaches, they are considered to involve the use of coercion. A third approach, the employment of psychiatric advance directives (Backlar, see chapter 10), a process that is for the most part still untried, may serve as a tool to help consumers develop self-management techniques, and provide a trigger to set in motion a PACT or OPC plan.

The Program for Assertive Community Treatment, aka the "Madison model of community care," was developed in the late 1960s (Santos, Hawkins, Julius, Deci, Hiers, & Burns, 1993; Stein & Test, 1980; Thompson, Griffith, & Leaf, 1990) in Madison, Wisconsin. The program was specifically designed for use with people who have serious mental disorders with psychotic features, who are unable to function adequately in the community between crises, and who repeatedly use crisis services, emergency rooms, and hospitals (Essock & Kontos, 1995). The assertive community treatment program operates much like a "hospital without walls." An interdisciplinary treatment team (psychiatrist, psychiatric nurse, social worker, case manager, etc.) provide direct treatment and rehabilitation continuously and without time limit to a specific client population wherever the clients live in the community. Programs that follow this model de-emphasize traditional office and facility based practice. As Drake and Burns (1995) describe, "the team is on call 24 hours a day for emergency treatment, provides home delivery of medications, actively monitors clients' physical health care, and has frequent contact with clients' family members." There is increasing evidence that assertive community treatment reduces the risk for psychiatric rehospitalization (Burns & Santos, 1995, McGrew, Bond, Dietzen, McKasson, & Miller; 1995). Approximately 34 states have such programs, but they are dissimilarly realized in different communities due to assorted social, political, and economic factors (Bachrach, 1988; Essock & Kantos, 1995). Programs that are more analogous to the prototype appear to be more effective in cutting down the number of days per year that a client spends in the hospital. Significantly, McGrew and associates' 1995 study showed that new assertive community treatment programs can be expeditiously implemented across an entire region of a state.

The prototype model, however, has been acknowledged to be paternalistic and coercive (Diamond & Wikler, 1985; Thompson, Griffith, & Leaf, 1990; Stein, 1990). A client's perceived "needs" trump his "wants." A reluctant client, perhaps unwilling to take medications, may be coerced into compliance by having her money allowance withheld. And if a client refuses assertive community services, willy-nilly, the services are likely to be continued. Paulson and colleagues (1999) have described a technique to reduce the tension associated with coercive outreach by utilizing other consumers as case management team members. These case managers, known as peer counselors, have an easier time establishing rapport and therefore may be more effective in the long run.

Another approach, involuntary outpatient commitment, is a patterned on the civil commitment model in which a court directs a person to comply with a particular *outpatient* treatment program. Thirty-five states and the District of Columbia provide for outpatient commitment by statute. However, even though findings from preliminary studies (Burns & Santos, 1995; McGrew, Bond, Dietzen, & Salyers, 1994) suggest that when used, treatment outcomes are improved and hospitalizations are decreased, outpatient commitment is used infrequently. And, if a state demands a person be judged dangerous before civil commitment can be obtained, outpatient commitment may be useless. When a person with "a serious mental illness has become dangerous, that person has often passed the point at which outpatient commitment is the appropriate placement alternative" (Torrey & Kaplan, 1995). A preventive commitment statute, which exists only in few states, authorizes outpatient commitment of individuals who do not, in the present, meet the customary commitment standard but would do so, precipitously, if intervention did not take place. Slobogin (1994) suggests this might be called a "predicted deterioration standard." The recent studies that identify the potential for violent behavior in persons with actively psychotic symptoms, may provide justification for such a standard.

Psychiatric advance directives, a third approach, has been little studied. An advance directive is a legal document that enables people to stipulate what medical treatment they wish to receive if, at a future time, they lose capacity to make decisions for themselves (Backlar, 1995a). Wexler (1994) envisages the potential for bringing psychological insights into the development of the law. He characterizes "therapeutic jurisprudence" as the law acting as a therapeutic agent to encourage the production of instruments, such as psychiatric advance directives, to facilitate consumers' authentic treatment choices. "A voluntary choice," according to Winnick (1994), "of a course of treatment involves a degree of internalized commitment to a goal often not present when the course of treatment is imposed involuntarily." Winnick argues that treatment, freely sought, is "more likely to be efficacious than treatment that is coerced." Appropriately fashioned and employed the advance directive may serve as a "voluntary" conduit to assertive community services or outpatient commitment for those who require such aid (Swanson, Tepper, Backlar, & Swartz, 2000). Psychiatric advance directives may be so designed through states' statutes, that a mentally ill person, who is decompensating, doesn't have to wait until her psychotic crisis is so severe that there is no alternative to hospitalization.

For those who have *capacity* to complete an advance directive for mental health treatment, this may prove to be the preferable path. But what about individuals who lack insight into their illness? A study by Amador and colleagues (1994) found that almost 60 percent of individuals with schizophrenia were moderately to severely incognizant of having a mental illness. Many such individuals are incapacitated and some are dangerous when not treated, but appear to function more or less adequately when treated (Geller, 1990). Will courts be sympathetic to their needs and legally sanction involuntary treatment, either inpatient or outpatient? Controversy, I suspect, about involuntary—coerced—treatment is likely to continue.

Appelbaum (1996) has noted that managed care and different ways of financing health care may bring about new patterns of care. With financial incentives that encourage the denial of care for persons with severe and chronic mental illness, managed care administrators are unlikely to attempt to force treatment upon those who refuse. Consequently, Appelbaum suggests that these issues may be debated in the realm of administrative law rather than standard mental health law. Now that we have a greater interest in how each health dollar is spent, society may no longer tolerate treatments that are inefficient. "The lack of long-term efficacy of many of the usual approaches to treatment of mental disorders, particularly chronic disorders, has been a loosely kept secret" (Appelbaum, 1996). Indeed, communities are already demanding that programs like outpatient commitment and assertive community treatment be used more comprehensively (Hernandez, 1999; Isaac & Jaffe, 1995; Stavis, 1999).

Will we find a way to avoid controlling and coercive care for some persons with mental disorders? In reality, none of us are free to control all aspects of our lives. Indeed, we control very little about our lives. Where we are born, the color of our skin, our temperaments, our abilities, all seem, for the most part, to be beyond our control. Likewise, people with mental disorders are controlled to a large extent by their bio-chemistry. It is easy to know, and perhaps see, if your arm is broken, but when the disease is in your brain it is hard to recognize what ails you. If you don't understand that you lack insight into your own affliction, it may be almost impossible for you to learn how to protect yourself—and others from yourself. Decompensating and delusional individuals usually are not intentionally violent or malicious but often are more despairing, fearful, and tormented than those whom they may beleaguer. We do not yet know how to cure persons with mental disorders; yet, as Callahan (1996, p. S12) observes, "healing may in a broader sense be possible even in those cases where medicine cannot cure. It can heal by helping a person cope effectively with permanent maladies."

Can we—as a society—justify dealing with a group of adult people, mentally ill people, differently than we deal with most others in our society? May we compel mentally ill people to take specific medications when we are unlikely to compel other adults to do the same? It would be an injustice, of course, to take an adverse action against any individual simply because he was a member of a specific population.

However, when statistical patterns identify a causal agent, like psychosis or alcohol that can make an individual act dangerously to others, to equivocate about offering some restorative therapy appears itself to be unjust. As Lefley (1996) observes "the 'normalization' of psychotic behavior by declaring it criminal rather than sick perpetuates the myth that mental illness is a myth and adds powerful reinforcement to its stigmatization."

Is there a *deus ex machina* that could have stopped the young man from killing his mother? Perhaps, like Maeterlinck's bluebird, the "gift from the gods" has been perched all the while on our very own mantelpiece. The gift, in this case, is the knowledge, supported by recent studies, that there is a connection, in some instances, found in a subgroup of mentally ill individuals, between violent behavior and mental disorder. Consumers, consumer advocates, families, and providers may have been afraid that by speaking this truth they would cast slurs and further stigmatize a population who historically have suffered dreadful indignities. Yet, to continue to sweep this information under the carpet "without a realistic look at dangerousness"—as Bloom (1989) perspicaciously has noted—is likely to advance such misconceptions.

## REFERENCES


Amador, X. F., Flaum, M., Andreasen, N. C., Strauss, D. H., Yale, S. A., Clark, C. C., & Gorman, J. M. (1994). Awareness of illness in schizophrenia and schizoaffective and mood disorders. *Archives of General Psychiatry, 51*, 826–836.

Appelbaum, P. S. (1996). Managed care and the next generation of mental health law. *Psychiatric Services, 47*, 27–28, 34.

Appelbaum, P. S., Lidz, C. W., & Meisel, A. (1987). *Informed consent: Legal theory and clinical practice*. New York: Oxford University Press.

Backlar, P. (1994). *The Family Face of Schizophrenia*. New York: Tarcher/Putnam.

Backlar, P. (1995a). The longing for order: Oregon's Advance Directive for Mental Health Treatment. *Community Mental Health Journal, 31*, 103–108.

Backlar, P. (1995b). Will the "Age of Bureaucracy" silence the rights versus needs debate? *Community Mental Health Journal, 31*, 201–206.

Bachrach, L. L. (1988). On exporting and importing model programs. *Hospital and Community Psychiatry, 39*, 1257–1258.

Beauchamp, T. L., & Childress, J. F. (2001). *Principles of Biomedical Ethics*, 5th ed. New York: Oxford University Press.

Bloom, J. D., (1989). The character of danger in psychiatric practice: Are the mentally ill dangerous? *Bulletin of the American Academy of Psychiatry and the Law, 17*, 241-255.

Binder, R. L., & MacNeil, D. E. (1986). Victims and families of violent psychiatric patients. *Bulletin of the American Academy of Psychiatry and the Law, 14*, 131–139.

Brown, P. (1985). *The Transfer of Care: Psychiatric Deinstitutionalization and its Aftermath*. London: Routledge & Kegan Paul.

Burns, B. J., & Santos, A.B. (1995). Assertive community treatment: An update of randomized trials. *Psychiatric Services, 46*, 669–675.

Butterfield, F. (1999, July 12). Prisons brim with mentally ill, study finds. *The New York Times*, p. A10.

Cook, J. A. (1988). Who "mothers" the chronically mentally ill? *Family Relations, 37*, 42–49.

Culver, C. M., & Gert, B. (1982). *Philosophy in Medicine: Conceptual and Ethical Issues in Medicine and Psychiatry*. New York: Oxford University Press.

Diamond, R. J., & Wikler, D. I. (1985). Ethical problems in community treatment of the chronically mentally ill. *New Directions for Mental Health Services, 26*, 85–93.

Drake, R. E., & Burns, B. J. (1995). Special section on assertive community treatment: An introduction. *Psychiatric Services, 46*, 667–668.

Essock, S. M., & Kontos, N. (1995). Implementing assertive community treatment teams. *Psychiatric Services, 46*, 679–683.

Estroff, S. E., Zimmer, C., Lachicotte, W. S., & Benoit, J. (1994). The influence of social networks and social support on violence by persons with serious mental illness. *Hospital and Community Psychiatry, 45*, 669–679.

Geller, J. L. (1990). Clinical guidelines for the use of involuntary outpatient treatment. *Hospital and Community Psychiatry, 41*, 749–755.

Grisso, T., & Appelbaum, P. S. (1992). Is it unethical to offer predictions of future violence? *Law and Human Behavior, 16*, 621–633.

Grob, G. N. (1991). *From Asylum to Community: Mental Health Policy in America*. Princeton: Princeton University Press.

Hernandez, R. (1999). Albany accord on confining of mentally ill. *The New York Times*, Wednesday, August 4, p. A17.

Hiday, V. A. (1992). Coercion in civil commitment: Process, preferences, and outcome. *International Journal of Law and Psychiatry, 15*, 359–377.

Hiday, V. A., Swartz, S. M., Swanson, J. W., Borum, R., & Wagner, H. R. (2001). Coercion in mental health care. In P. Backlar, & D. L. Cutler (Eds.), *Ethics in Community Mental Health Care: Commonplace Concerns*. New York: Kluwer Academic/Plenum Publishers.

Isaac, R. J., & Armat, V. C. (1990). *Madness in the Streets: How Psychiatry and the Law Abandonded the Mentally Ill*. New York: The Free Press.

Isaac, R. J., & Jaffe, D. J. (1995). Mental illness, public safety. *The New York Times*, December 23, p. A17.

Junginger, J. (1995). Command hallucinations and predictions of dangerousness. *Psychiatric Services, 46*, 911–913.

Lefley, H. (1987). Aging parents as caregivers of adult mentally ill children. *Hospital and Community Psychiatry, 38*, 1063–1070.

Lefley, H. (1989). Family burden and family stigma in major mental illness. *American Psychologist, 44*, 556–560.

Lefley, H. (1996). *Family Caregiving in Mental Illness*. Thousand Oaks, California: Sage Publications.

Link, B. G., Andrews, H., & Cullen, F. T. (1992). Reconsidering the violent and illegal behavior of mental patients. *American Sociological Review, 57*, 275–292.

Macklin, R., (1993). *Enemies of Patients*. New York: Oxford University Press.

McGrew, J. H., Bond, G. R., Dietzen, L., & Salyers, M. (1994). Measuring the fidelity of implementation of a mental health program model. *Journal of Consulting and Clinical Psychology, 62*, 670–678.

McGrew, J. H., Bond, G. R., Dietzen, L., McKasson, M., & Miller, L. (1995). A multisite study of client outcomes in Assertive Community Treatment. *Psychiatric Services, 46*, 696–701.

Mill, J. S., ([1859] 1977). *On liberty*. In J. M. Robson (Ed.), *Collected works of John Stuart Mill*, vol. *XVIII*. Toronto: Toronto University Press.
Monahan, J. (1992a). "A terror to their neighbors": Beliefs about mental disorder and violence in historical and cultural perspective. *Bulletin of the American Academy of Psychiatry and the Law*.
Monahan, J. (1992b). Mental disorder and violent behavior: Perceptions and evidence. *American Psychologist*, *47*: 511–521
Mulvey, E. P. (1994). Assessing the evidence of a link between mental illness and violence. *Hospital and Community Psychiatry*, *45*, 663–668.
Murphy, J. (1976). Psychiatric labeling in cross-cultural perspective: similar kinds of disturbed behavior appear to be labeled abnormal in diverse cultures. *Science*, *191*, 1019–1028.
Newhill, C. E., Mulvey, E. P., & Lidz, C. W. (1995). Characteristics of violence in the community by female patients seen in a psychiatric emergency service. *Psychiatric Services*, *46*, 785–789.
Paulson, R., Herinckx, H., Demmler, J., Clarke, G., Cutler, D., & Birecree, E. (1999). Comparing practice patterns of consumer and non-consumer mental health service providers. *Community Mental Health Journal*, *35* (3), 251–269.
Rawls, J. (1971). *A Theory of Justice*. Cambridge, Massachusetts: Harvard University Press, p. 442.
Rothman, D. J. (1987). Ethics and human experimentation: Henry Beecher revisited. *New England Journal of Medicine*, *317*, 1195–1199.
Rothman, D. J. (1991). *Strangers at the Bedside: A History of How Law and Bioethics Transformed Medical Decision Making*. New York: Basic Books.
Santos, A. B., Hawkins, G. D., Julius, B., Deci, P. A., Hiers, T. H., & Burns, B. J., (1993). A pilot study of assertive community treatment for patients with chronic psychotic disorders. *American Journal of Psychiatry*, *150*, 501–504.
Scull, A. (1989). *Social Order/Mental Disorder: Anglo-American Psychiatry in Historical Perspective*. Berkeley: University of California Press.
Shah, S. A. (1990). Violence and the mentally ill. *The Journal of the California Alliance for the Mentally Ill*, *2*, 20–21.
Slobogin, C. (1994). Involuntary community treatment of people who are violent and mentally ill: A legal analysis. *Hospital and Community Psychiatry*, *45*, 685–689.
Stavis, P. (August 19, 1999). Treatment by cooperation: Taking exception. *Washington Post*, Op Ed Page.
Stein, L. I. (1990). Comments by Leonard Stein. *Hospital and Community Psychiatry*, *41*, 649–651.
Stein, L. I., & Test, M. A. (1980). Alternative to mental hospital treatment: I. conceptual model, treatment program, and clinical evaluation. *Archives of General Psychiatry*, *37*, 392–397.
Steinwachs, D. M., Kasper, J. D., & Skinner, E. A. (1992). *Family perspectives on meeting the needs for care of severely mentally ill relatives. A national survey*. Arlington, Virginia: National Alliance for the Mentally Ill.
Straznickas, K. A., McNeil, D. E., & Binder, R. L. (1993). Violence toward family caregivers by mentally ill relatives. *Hospital and Community Psychiatry*, *44*, 385–387.
Swanson, J., Holzer, C., Gnaju, V., & Jono, R. (1990). Violence and psychiatric disorder in the community: Evidence from the Epidemiologic catchment area surveys. *Hospital and Community Psychiatry*, *41*, 761–770.
Swanson, J. W., Tepper, M. C., Backlar, P., & Swartz, M. S. (2000). Psychiatric advance directives: An alternative to coercive treatment? *Psychiatry*, *63*, 160–172.

Tax lawyer's son committed. *The New York Times*, May 9, 1983, p. 15.

Thompson, S. T., Griffith, E. E. H., & Leaf, P. J. (1990). A historical review of the Madison model of community care. *Hospital and Community Psychiatry*, *41*, 625–633.

Torrey, E. F. (1994). Violent behavior by individuals with serious mental illness. *Hospital and Community Psychiatry*, *45*(7), 653–662.

Torrey, E. F., & Kaplan, R. J. (1995). A national survey of the use of outpatient commitment. *Psychiatric Services*, *46*, 778–784.

Wexler, D. B. (1994). An orientation to therapeutic jurisprudence. *The New England Journal on Criminal and Civil Confinement*, *20*, 259–264.

Winick, B. J. (1994). The right to refuse mental health treatment: A therapeutic jurisprudence analysis. *International Journal of Law and Psychiatry*, *17*, 99–117.

CHAPTER 6

# AT WORK WITH THREATS AND VIOLENCE

CARL C. BELL, MORRIS A. BLOUNT, JR., AND TANYA R. ANDERSON

There is a great deal of ambivalence in the literature regarding violence perpetrated by the mentally disordered. Early studies indicated psychiatric patients were less violent than the general population (Ashley, 1922; Pollock, 1938; Cohen & Freeman, 1945), but subsequent investigations found opposite findings (Giovannoni & Gurel, 1967; Lagos, Perlmutter, & Saexinger, 1977; Rappaport & Lassen, 1965; Sosowky, 1978; Zitrin, Hardesty, Burdock, & Drossman, 1976). Steadman, Cocozza, & Melick (1978) suggested the change over time was due to the increased numbers of patients with criminal records entering psychiatric facilities. Initially, Monahan and Steadman (1983) concluded, if a number of sociodemographic factors related to crime are taken into account, the correlation between mental illness and criminality was weak. Similarly, Teplin (1985) asserted the mentally disordered did not constitute a dangerous group prone to violent crime based on actual observations of police-citizen interactions. More recently, Monahan (1992) has reconsidered his position and now believes there is a relationship between mental disorder and violent behaviors. Additionally, Mulvey (1994) notes there is a connection between mental illness and violent behavior in the community.

Recently, Marzuk (1996) pointed out many flaws in this earlier research citing the lack of consistent definitions of violence and/or mental illness, the reliance on archival data, the lack of control of demographic and situational variables, and the retrospective nature of the studies. Backlar (in press) admirably points out the various factors (e.g. the mentally disordered not being available in the community to be arrested prior to deinstitutionalization and the wish not to stigmatize the mentally disordered), which may have also accounted for the conflicting perspectives these studies represented. More convincing recent studies have improved on earlier study designs by examining large community samples (Swanson, 1994). Link, Andrews, and Cullen (1992) have also teased out the differentiation between trait and state violence, and suggest that mentally ill may only be more violent than the general population when in the throes of their psychosis.

Despite the accuracy of these observations, the controversy about the nature of the relationship between violence and mental illness remains. For example, Link, Andrews, and Cullen can't say for sure if the psychotic symptoms cause violence or if it is the general belief that mentally ill are more violent causing reactions to the mentally ill that triggers violence. Regardless of the facts, as Backlar (in press) points out, the reality is there is a belief that mentally ill patients are more likely to be dangerous, and this belief is held by other patients, family, and mental health professionals alike.

Accordingly, anyone coming into contact with a patient whose behavior is perceived to be unpredictable is going to feel at risk for being victimized by that patient. Thus, patients, family members, and mental health practitioners are all in the same boat. Having worked in a psychiatry emergency room for several years where 20% of the patients presented in an agitated manner, and 15.6% of the total patient population had to be restrained for out of control behavior (Bell & Palmer, 1981). We can only imagine the fear and anxiety that patients and family members feel (Backlar, 1994) when a mentally ill patient who is in their midst continuously isn't stable on medication and acting strangely.

Despite not actually living with the mentally ill, the issue of violence in the mentally ill is a serious one for those of us on the front lines treating patients in the community. While the proposition that appropriately treated patients are as violent as the average citizen is true, one issue which is rarely discussed is the personal safety of the clinician and how clinicians should respond to aggression. When confronted with an aggressive, potentially violent patient, personal safety should always be a primary concern. Unfortunately, because of clinician experiences and attitudes, this concern may not be manifest. One point to remember, however, is though a clinician's mission is to save lives, lessen harm and do no harm, the healer must himself be safe from harm. (Berg, Bell, & Tupin, 2000). It has always been clear to me us an appropriate and ethical response to an actual physical attack by a patient is self-defense. For example if a 95-year-old man with a brain tumor attacks with a knife in an effort to kill a clinician, he or she has the right to pick up a chair and hit him with it in an effort to preserve his or her life. Of course, if the clinician has knocked him unconscious, but continues to hit a defenseless patient because the clinician is mad about the attack, then he or she should be put in jail for assault and battery. In addition to this last resort intervention of self-defense, the clinician should be skilled in knowing how to identify patients who are on the verge of becoming overtly violent and should have social skills which would allow him or her to verbally de-escalate the patient's near violent disposition (Bell & Jenkins, 1995).

We suspect because frank discussions about the issue of self-defense are not held in training, there is a fair amount of ambivalence and anxiety around the issue of personal safety with the result being the quandary is rarely adequately addressed. We wonder if it is ethical for the profession to continue to allow this discussion to go unattended, as it is obvious to us until clinicians feel safe they will continue to avoid assessing and treating patients who are potentially violent. Further, when it comes to addressing issues of violence with patients or family members, this topic will be actively avoided with the result being the omission of planning for violent behavior. When violence does occur, everyone, the clinician, the patient, and the family will be unprepared.

Another problem is the lack of clarity around when to hold a patient responsible for his or her violent behavior. It is plain to us there are some patients in the throes of an acute psychotic disorders who, despite being violent, should not be held responsible for their behavior. In addition, there are some patients who, although suffering from various forms of mental illness, use violence as a tool to further their predatory aims and who should be held completely responsible for their behavior, even if this behavior occurs in a treatment setting, such as an inpatient unit or outpatient clinic.

I am reminded of two incidents that occurred while doing an inpatient rotation during residency training (MB). Both events occurred within weeks of each other and both involved patients with chronic psychotic disorders. Though the incidents appeared similar on the surface, there were important differences in circumstances that therefore resulted in different outcomes. The first situation involved a patient in her early twenties who had been on the unit for a few days and had already been noted to challenge authority, usually by objecting to various floor rules. On this particular occasion she was being redirected from the dayroom because of an upcoming meeting. After being told by one staff member she could not go to the day room at that time she ran up to an uninvolved staff member and began hitting her. While she was being contained, she threatened to have gang-associated acquaintances of hers kill a couple of members of the nursing staff and stated she wanted revenge for the staff member telling her what to do. As a result, she had legal charges pressed against her. She later said she had been violent before to staff members at other facilities but had never had even a threat of legal action being taken against her; she was surprised the staff was willing to file charges and hold her accountable for the actions for which she was aware were wrong. The second scenario also involved a woman in her twenties who had been on the unit for a few weeks. One morning she ran down the hallway to an unsuspecting medical student and began punching the student without apparent provocation. She had no recollection of her actions and was later found to have interictal psychosis in addition to her paranoid-producing psychotic illness. Because it was presumed she was actively psychotic at the time, the legal system was not involved and she was treated accordingly.

The response to the first type of behavior is to use a legal/criminal justice approach while the response to the second type of behavior is treatment. Unfortunately the delineation between these two circumstances of violence is poorly conceptualized and the reaction to these two situations is often ineffectual. This delineation is even more complicated when the patient is an adolescent and the legal involvement is with the juvenile justice system. Though the juvenile justice system is moving closer to the principles of the adult legal model and away from its origin in rehabilitation, remnants of the old philosophy still permeate daily practice. As such, many juvenile court justices are even less likely to adjudicate youth for violent behavior. Adolescents quickly realize that there may be little or no consequences for their behavior and the behavior may be exacerbated. This deficiency between the functioning of the two systems inadvertently supports the adolescent's feeling of invulnerability and may retard or disrupt some normal adolescent developmental processes. This may lead to continued functional difficulties in adulthood. The adolescent may or may not exhibit a true psychiatric disorder, which may or may not receive appropriate treatment. We feel anyone with a history of mental illness who perpetrates violence needs an evaluation regarding the motivation for his or her behavior and should be managed according to his or her intent.

I (CB) recall seeing a patient who threatened to go home, get his gun, come back to the clinic, and kill me because I refused to give him four 25 milligram tablets of chlorpromazine to take at bedtime as needed instead of a one 100 milligram tablet to take at bedtime regularly. During the visit prior to his threatening visit, we had discussed how he had been getting his requested regimen of four 25 milligram tablets of chlorpromazine for several months, but, since he never took the full one hundred milligrams, he would frequently come to the day treatment program in an agitated, irritable, impulsive fashion. During this visit prior to his threat we made a compromise agreement—we would try him on the four 25 milligram tablets for one more month. If he only took one or two or three of the 25 milligram tablets at bedtime as he felt necessary, and if it was apparent to the day treatment staff that he continued to be impulsive, irritable, and agitated; then he would agree to take a one 100 milligram tablet regularly at bedtime as this regimen had helped to control his problematic behavior in the past.

During the visit of his threat, I reminded him of our agreement the month before, and, since he had not done well taking the 25 milligram tablets, I invoked our agreement and prescribed the 100 milligram tablets. My holding him to our agreement and my refusal to give him the medication the way he wanted it caused him to make his death threat and run out of the clinic. Taking him quite seriously, I called the police, and attempted to press charges. Three hours later, after talking to the two original policemen who arrived on the scene, then their sergeant, and then his lieutenant, I explained to them, while the patient had a mental illness, he was not currently overtly psychotic and therefore not a candidate for certification to a state hospital for dangerousness secondary to a mental illness. I finally got a warrant made out for the patient's arrest for making a criminal threat against me.

A few days later, after waiting six hours in court, I got the opportunity to spend 15 minutes insisting the judge take this issue seriously and not dismiss it. Apparently, the police had not acted on the warrant and the patient was not present in court. After several minutes of the judge questioning my competence as a psychiatrist for not dropping the charges he issued a continuance, and I was referred to the police for an order of protection. Of course they tried to talk me out of filing the order. The next day the patient, an ex-Viet Nam veteran, called the clinic and threatened to blow it up in an effort to kill me. I turned this information over to the police, but they didn't seem impressed. When I went to court the second time and the patient did not show up, rather than have another futile continuance, I finally gave up.

All through this process I had consulted with the Department of Mental Health on how to proceed because I had questions about issues of confidentiality and wondered about my obligations to provide treatment to a patient who seemed to want to harm me due to his malicious intent. The Department of Mental Health was of little help other than to insist that if I banned the patient from the agency, the patient needed to be referred elsewhere for treatment. So, we sent him a letter banning him from the agency along with a referral to another agency, which agreed to serve him. About three days later the windows on the residential van were broken out and the next day, the patient called and admitted to doing it to get even. I hoped the vengeance on the agency vans would assuage his need to kill me. A few days after he broke out the van windows; our state hospital liaison discovered him in the hospital where she made it understandable to him that he could not come back to the agency, and when discharged, he would be linked to the agency to which we had referred him.

Our experience is when mentally ill patients seriously threaten staff or assault staff due to predatory intent it is best to file criminal charges against them and refer them for treatment to another agency with the understanding that the new agency will review the reason for the referral with the patient and make it clear to them that such behavior will not be tolerated. This intervention sends a clear message to the potentially violent patients with predatory or bullying intent, and curtails their future use of violence as an instrument to achieve their goals. On the other hand patients who are violent towards staff due to an acute psychotic process do best when they are welcomed back into the center after they are in remission from their psychotic process and after they have had a session clarifying what happened during their violence, why it happened, and how to prevent it from happening again.

Backlar (in press) appropriately raised the issues of the ethics of coercion contained in involuntary outpatient commitment and assertive community treatment. We think we have a very long way to go to swing the pendulum back to a reasonable point. The civil libertarians and anti-psychiatrists who have tried to dismantle the process of involuntary psychiatric treatment have gone too far. There have been too many times when we have attempted to hospitalize a patient who was gravely disabled only to have our efforts frustrated by the way the current legal system over protects a seriously mentally ill patient's "rights". Our frustration has been compounded when the same patient is brought back to the emergency room a week later in even worse shape or harmed by his or her own behavior. We are not in favor of paternalism or coercion unless it is empirically clear, based on a lengthy pattern of predictable behavior that a patient is likely to be of harm to himself by virtue of his mental illness not getting treated.

I (CB) recall a patient whom I saw regularly in the emergency room who would usually be brought against his will for treatment due to being gravely disabled by virtue of his schizophrenic illness. I repeatedly certified this patient to the state hospital where he would eventually be committed for treatment. As soon as he was sufficiently recovered from his acute symptoms and able to live in the community he would be released. Unfortunately and predictably, he would return to the emergency room after a couple of months, as he would not comply with the outpatient treatment that had been recommended on his discharge and we would repeat the cycle. Finally, after six or seven cycles, I began to treat him in the psychiatric emergency room with rapid tranquilization and a long acting neuroleptic medication, which would allow him to recover without needing admission to a state facility (Bell, 1978). Of course he continued on his cycle of never coming to see me on an outpatient basis and returning back and forth to the psychiatric emergency room for treatment. Finally, during one of his psychotic episodes the police caught him stealing a purse and a condition of his probation was he get outpatient mental health care for nine months along with taking medication to treat his illness.

He began to come regularly for outpatient treatment and I began to see him in the clinic every two weeks to administer a long-acting neuroleptic medication. The improvement in the patient's behavior and lifestyle was dramatic. Prior to his regular treatment in the clinic, when he would have acute psychotic relapses he would, with some regularity, assault his father. With routine outpatient treatment and medication this violence stopped. Prior to his regular treatment in the clinic when he was brought to the emergency room by the police, he would smell of feces; when he was ill he was so regressed he would not use the toilet and soil himself. After his treatment, his white undershorts, which had heretofore been a dark brown, returned to their original white color due to his vastly improved toilet habits and hygiene. He also was able to move from his father's home and live independently in addition to deciding to finish his college education. By the patient's own independent evaluation things were going exceedingly well since he had been coerced into treatment, for he used to talk at length about the suffering he encountered while acutely ill. Unfortunately, his lack of insight prohibited him from making the connection between the treatment he was receiving and his improvement in psychosocial functioning.

When the probationary period ended, despite my protestations, he chose to stop coming to treatment because he didn't think it was necessary and was lost to follow-up. I could not counteract his decision as he was no longer obligated to see me and he was completely rational despite not having insight into the nature and pattern of his schizophrenic illness. A few months later, I learned from his father the patient had, while in control of his faculties, moved outside of the state. He deteriorated again and the night before his suicide the patient had called his father to lament his suffering from his mental illness before ending his life. Clearly, for patients like this, who have a well-established pattern of floating in and out of competency due to an inability to understand the connection between their competency and treatment and a tendency to be violent when acutely psychotic, some form of coercion should be required.

Although technically not involuntary outpatient commitment (i.e. after the patient committed a crime he was ordered to treatment by virtue of being a condition of probation), coercion worked in this patient's case, and we suspect in such instances involuntary outpatient commitment would work. A tactic to satisfy the civil libertarians might be to wait until a mentally ill person commits a crime, e.g. assault on a family member, and then prosecute the patient with the understanding, rather than spend time in jail for his or her crime, he or she would be required to go to treatment for a reasonable length of time, e.g. a year after the first offense, two years after the second offense, and five years after the third offense. While it would be better to have the ability to predict dangerousness due to psychiatric illness, and forego a patient actually having to harm someone before he or she could be coerced into treatment for some predetermined length of time, such a system would be an improvement over the one which is presently in operation. Unfortunately, although Backlar (in press) suspects that "predictive testimony about future violence may have sufficient scientific support and may become more acceptable to psychiatry, psychology, and the courts," reliable predictions about potential violence, which will be strong enough to call for coercive action to force patients into community treatment, are a long way off in the future.

Currently, what happens is a patient will assault a family member while mentally ill and the police, based on the family's information that the perpetrator is mentally ill, will bring the patient to the emergency room. In these instances the police always insist on the family member dropping criminal charges, and request the patient get certified to the psychiatric hospital—only for the patient to get out in 2–3 weeks without any assurance that the patient will get treatment. Pressing criminal charges for a mentally ill patient's attack (in the hopes of getting treatment mandated as a condition of probation) may cause the patient, who should not be held responsible for their behavior, to spend time in jail for their crime. This is already occurring with some alarming frequency because sometimes acutely ill patients will break the law, but their behavior is not recognized as a product of their mental illness.

If the patient is lucky to get treatment while in jail, the chances that he or she will be adequately linked to an outpatient treatment facility after are release from jail may actually be less than if they had been hospitalized in an inpatient psychiatric unit. They now bear the label of being a criminal and some mental health centers are reluctant to accept such patients as they think their violence may have been predatory in nature rather than being a product of mental illness. For patients who do use violence as an instrument to bully their families into submission, the possibility of them going to jail like everyone else who uses violence in a predatory manner might dissuade them from their intentions as it does for most of us in society.

We believe involuntary outpatient commitments should be used sparingly in appropriate cases after proper safeguards have been applied as we don't agree with some advocates who seem to prefer a blanket use of involuntary outpatient commitment for patients who drink, are mentally ill, and have a past history of violence (distant or otherwise) (Torrey, 1994). We more tend to agree that patients who exhibit persistent violence (i.e., trait violence) should be certified to outpatient treatment (Slobogin, 1994). More information, however, needs to be obtained about specific factors which may be associated with increased violence in mentally ill patients. For example, Swanson et al. (2000) have found certain baseline risk factors to be associated with increased violence. Some of these risk factors include: being under 40 years of age, being single, having low levels of social support, living in an urban area, being recently homeless, having a GAF of 47 or below, and being a substance abuser. Other factors include having psychotic symptoms that involve perceived threat and an override of internal cognitive control, having paranoid symptoms, and experiencing more than two hospitalizations within the previous year. Perhaps evidence such as this can be used to strengthen the predictive testimony about future violence to protect health care providers, family members, and patients themselves.

We agree with Backlar's doubt we can avoid coercive care for some persons with mental disorders (in press), and would offer the following experience which suggests assertive community treatment without an element of coercion cannot succeed. Illinois tried to institute a type of assertive community treatment program for a group of patients who had had at least three hospitalizations within the past year, but which was not coercive in nature as the patient had complete autonomy about refusing treatment without any consequences (Illinois Department of Mental Health and Developmental Disabilities, 1990). I (CB) can still recall a gripping story of one of the case managers spending considerable time with a patient who, in addition to being mentally ill, liked to spend her disability check for drugs. As a result, she would run out of rent and get evicted from her apartment. Since the facility was prohibited from being coercive, the staff attempted to secure housing for her from the funds they were getting to take care of her. Of course the capitated rate for the patient's care was to meet all of her needs, but it wasn't enough to cover her needs once she had squandered her disability check.

I can still recall the case manager complaining about dodging traffic (the patient had a penchant for standing in the middle of a busy intersection and directing traffic while high) while trying to talk the patient into spending the night in the nice apartment he had guaranteed for her instead of sleeping on a park bench in the cold where she had been previously assaulted. The program was referred to as the CILA (Community Integrated Living Arrangement) Program, but was quickly renamed the "SILLY" Program by staff as it was an oxymoron to expect historically uncooperative, uninsightful patients to agree with treatment when they had been given free reign to thwart all of our efforts. The technical legal procedures aren't as important to us as long as the determination to restrict a person's rights based on their having a mental illness is carefully done and not over applied. Recently, there has been a chronically psychotic patient in clinic (MB) who had been non-compliant with treatment, coming into the clinic only to receive his weekly allowance; the clinic serves as his payee. When he is not being treated with medications he is paranoid towards his family, refuses treatment and does not have personal hygiene, for he does not bathe or change clothes for weeks at a time. Though he has a supportive family, he chooses to live on the streets, reportedly eating from and sleeping around garbage dumpsters. He also occasionally gets into altercations with others because of his paranoia, though never taken in by the police because of his psychosis. Until recently, he would come into the clinic regularly to pick up his money but would refuse to stop by my office to receive his long-acting neuroleptic shot. After discussion with the treatment team, we decided to be more aggressive about him attending the appointments with me. He was told his treatment required visiting the doctor before receiving his allowance. For the first few weeks after being told this he stopped by my office weekly but still refused medication. He would stay long enough to say he did not need medication. During these visits he was reminded that he once took shots and they were a part of his treatment. At the same time his family persuaded him to comply with treatment whenever they saw him. Well, after being reminded that he received the long-acting neuroleptic in the past without negative side effects to him and after strongly implying he needed to receive the shot to receive his allowance, he finally agreed to "try one shot." After a few weeks of this he became less paranoid and actually began checking-in with me to confirm his next appointment for injection of the long-acting neuroleptic. He also has made remarkable clinical improvement. Though this patient had only occasional violent outbursts, the coercion, if viewed that way, has benefited him clinically.

Presently, there are patients who are violent secondary to having an active mental illness or who are violent due their predatory intent. They are either not being claimed by the mental health system or the criminal justice system or are in the wrong system. There are patients who are violent as a result of their mental illness who are in the criminal justice system, and there are patients who are more predatory than actively mentally ill who are in the mental health system. Hopefully, most of the patients wind up in the right place, but many are in the wrong system or float back and forth between the two systems not fully belonging to either system. To adequately respond to the issue of threats and violence we must become more sophisticated in differentiating the various motives for an individual's violent behavior. We must also balance the different perspectives the patients and family of patients bring to the table regarding coerced treatment; they both have some valid positions (Bell, 1994).

The idea of advanced consent proposed by Backlar (in press) is an interesting one; at the Community Mental Health Council we use this technique to cope with issues of releasing information. Frequently, when patients are intact they are quite willing to allow us to communicate their progress with their family members. Sometimes when patients decompensate they become paranoid about family members, and when the family begins to notice their loved one's deterioration and requests information, the patient (in their paranoid state) refuses to allow the therapist to inform the family about what is happening to the patient. By having advanced consent encompassing this contingency we are in a better position to serve the patient's needs.

Backlar's (this volume) proposition that society would rather adequately treat to prevent violence rather than place violent patients in the correctional system is a bit naive to us as we reflect on how society responded to a medical illness called drug addiction. Despite this danger, I agree with her advocacy which encourages patients, family members, and mental health professionals to begin to own up to "the fact (supported by recent studies) that there is a connection in some instances, found in a sub-group of mentally ill individuals, between violent behavior and mental disorder." However, we have a lot of careful work to do on this issue before we will feel comfortable that we are doing our best.

## REFERENCES


Ashley, M. (1922). Outcome of 1,000 cases paroled from the Middletown state homeopathic hospital. *New York State Hospital Quarterly, 8*, 64–70.

Backlar, P. (1994). *The family face of Schizophrenia* (p. 14). New York: Tarcher/Putnam.

Backlar, P. (2001). At home with threats and violence. In: P. Backlar, & D. Cutler (Eds.), *Ethics in community mental health care: Commonplace concerns*. New York: Kluwer Academic/Plenum Publishers.

Bell, C. C. (1978). The role of psychiatric emergency services in aiding community alternatives to hospitalization in an inner city population. *Journal of the National Medical Association, 70*(12), 931–935.

Bell, C. C. (1994). Violent behavior and mental illness: Perspectives on papers in this issue. *Hospital and Community Psychiatry, 45*(7), 711–713.

Bell, C. C., & Jenkins, E. J. (1995). Violence prevention and intervention in juvenile detention and correctional facilities. *Journal of Correctional Health Care, 2*(1), 17–38.

Bell, C. C., & Palmer, J. (1981). Security procedures in a psychiatric emergency service. *Journal of the National Medical Association, 73*(9), 835–842.

Berg, A. Z., Bell, C. C., & Tupin, J. (2000). Clinical safety: Assessing and managing the violent patient. *Psychiatric aspects of violence: Issues in prevention and treatment*, C. C. Bell, (Ed.), 2, pp. 9–29.

Cohen, L., & Freeman, H. (1945). How dangerous to the community are state hospital patients, *Connecticut State Medical Journal, 9*, 697–700.

Giovannoni, J. M., & Gurel, L. (1967). Socially disruptive behavior of ex-mental patients. *Archives of General Psychiatry, 17*, 146–153.

Illinois Department of Mental Health and Developmental Disabilities, (1990). "Community-Integrated living arrangements: People, choices, new beginnings." Springfield, IL: Illinois Department of Mental Health and Developmental Disabilities.

Lagos, J., Perlmutter, J., & Saexinger, H. (1977). Fear of the mentally ill: Empirical support for the common man's response. *American Journal of Psychiatry, 134*, 1134–1137.

Link, B. G., Andrews, H., & Cullen, F. T. (1992). The violent and illegal behavior of mentally patients reconsidered. *American Sociologic Review, 57*, 275–292.

Marzuk, P. M. (1996). Violence, crime, and mental illness. *Archives of General Psychiatry, 53*, 481–486.

Monahan, J. (1992). Mental disorder and violent behavior: Perceptions and evidence. *American Psychologist, 47*, 511–521.

Monahan, J., & Steadman, H. (1983). Crime and mental disorder: an epidemiologic approach. In: M. Tonry, & N. Morris (Eds.). *Crime and justice: An annual review of research*. Chicago: University of Chicago Press, pp. 145–189.

Mulvey, E. P. (1994). Assessing the evidence of a link between mental illness and violence. *Hospital and Community Psychiatry, 45*(7), 663–668.

Pollock, H. (1938). Is the paroled patient a menace to the community? *Psychiatric Quarterly, 12*, 236–244.

Rappaport, J. R., & Lassen, G. (1965). Dangerousness-arrest rate comparisons of discharged patients and the general population. *American Journal of Psychiatry, 121*, 776–783.

Slobogin, C. (1994). Involuntary community treatment of people who are violent and mentally ill: A legal analysis. *Hospital and Community Psychiatry, 45*(7), 685–689.

Sosowky, L. (1978). Crime and violence among mental patients reconsidered in view of the new legal relationship between the state and the mentally ill. *American Journal of Psychiatry, 135*(1), 33–42.

Steadman, H. J., Cocozza, J. J., & Melick, M. E. (1978). Explaining the increased arrest rate among mental patients: The changing clientele of state hospitals. *American Journal of Psychiatry, 135*, 816–820.

Swanson, J. (1994). Mental disorder, substance abuse, and community violence. In J. Monahan, & H. J. Steadman (Eds.). *Violence and mental disorders: Developments in risk assessment*. Chicago: University of Chicago Press.

Swanson, J., Swartz, M., Borum, R., Hiday, V., Wagner, H., & Burns, B. (2000). Involuntary out-patient commitment and reduction of violent behavior in persons with severe mental illness. *British Journal of Psychiatry, 176*, 324–331.

Teplin, L. A. (1985). The criminality of the mentally ill: A dangerous misconception. *American Journal of Psychiatry, 142*(5), 593–599.

Torrey, E. F. (1994). Violent behavior by individuals with serious mental illness. *Hospital and Community Psychiatry, 45*(7), 778–781.

Zitrin, A., Hardesty, A. S., Burdock, E. I., & Drossman, A. K. (1976). Crime and violence among mentally ill patients. *American Journal of Psychiatry, 133*(2), 142–149.

# Part IV

# INVOLUNTARY INTERVENTIONS

CHAPTER 7

# INVOLUNTARY INTERVENTIONS IN DUAL DISORDERS PROGRAMS

DOUGLAS L. NOORDSY, CAROLYN C. MERCER, AND ROBERT E. DRAKE

In this chapter, we discuss the moral and clinical evaluation of involuntary interventions in the treatment of clients with dual disorders. Many techniques are available for assisting clients to engage and comply voluntarily in dual disorders treatment. These techniques include motivational interviewing, treatment plan contracts, informal agreements, and behavioral contracts. Like other clients with severe mental illnesses, clients with dual disorders may sometimes experience involuntary intervention because they appear likely to harm themselves or others. Indeed, clients with dual disorders are at particular risk for behaviors that might harm themselves or others (e.g., see Bartels, Drake, Wallach, & Freeman, 1991; Cuffel, 1994; Cuffel, Shumway, & Chouljian, 1994; Lindqvist & Allebeck, 1994; Yesavage & Zarcone, 1983). Involuntary interventions by definition restrict personal choice and freedom. Such interventions take away the client's self-control and can undermine the client's self-motivation. Since self-motivation and self-control are strengths that dual disorders treatment aims to cultivate, the use of involuntary interventions requires particular care with clients who have dual disorders.

## BACKGROUND

Involuntary interventions raise a number of legal, political, ethical, moral, and clinical questions that are important to both administrators and clinicians; guidance for mental health clinicians has been published elsewhere (e.g., American Psychiatric Association, 1987, 1993; Group for the Advancement of Psychiatry, 1994; Monahan, Hoge, Lidz, Roth, Bennett, Gardner, & Mulvey, 1995; Mulvey, Geller, & Roth, 1987). The legal mandate for involuntary intervention is essentially a paternalistic action by and on behalf of society. Medical ethicists Culver and Gert (1982) have written that a paternalistic action is justified if certain moral rules are fulfilled, or if a client consents voluntarily to the action. The moral rules governing a paternalistic action are fulfilled if:

1. The harm to the client that would be prevented by the paternalistic action is so significant that the client could not rationally prefer the alternative; and
2. The client does not have an adequate reason for suffering the harm that the paternalistic act is designed to prevent; and
3. The harm is significantly likely to occur and the paternalistic act will significantly diminish the probability of its occurrence.

Justification therefore hinges on: (1) the seriousness of harm to be avoided; (2) the extent of person's rational responsibility; (3) the likelihood of harm, and (4) the likelihood that the involuntary intervention will significantly diminish the likelihood of harm.

In mental health care, involuntary interventions are legally mandated mechanisms for limiting a person's choices, and the specifics vary from state to state. The mechanisms fall into three categories—protections for person and property, mandates for treatment, and orders associated with illegal behaviors. *Protections for person and property* take the form of payeeships for Supplemental Security Income, Social Security Disability Insurance, and other funds, as well as guardianships or conservatorships over the person, the person's medical care, or the person's property. *Mandates for treatment* include orders for emergency inpatient or outpatient treatment, involuntary hospitalization, conditional discharge from the hospital, and inpatient or outpatient commitments. *Orders associated with illegal behaviors*, which can be tailored to address dual disorders treatment goals, encompass restraining orders, conditions for avoiding prosecution, detention, conditions of sentencing, and stipulations or conditions associated with probation or parole.

Outside the mental health system, involuntary interventions are frequently used in substance abuse treatment where a person's addiction has placed others in jeopardy. Indeed, many substance abuse treatments are coercive in the sense that family members, employers, courts, police, or others are pushing the client into treatment. In the mental health system, involuntary interventions are widely sanctioned by laws, but are quite negatively regarded by clients and client advocates. Most clinicians and administrators are reluctant to use involuntary interventions except when necessary to prevent serious harm.

The laws of every state sanction the coercion of people with mental illnesses into treatment, usually with the requirement that the person would be likely to harm himself or others. The actual use of involuntary hospitalization varies from place to place. While 27% of psychiatric hospital admissions are involuntary across all states, the rate is as high as 75% in some states (Monahan et al., 1995). Procedures and criteria vary from state to state as well, with some states having separate procedures for brief emergency commitment, for longer observational commitment, and for longer-term extended commitment. Research efforts are clouded by these differences and are also made complex by the different ways that clients perceive involuntary status and coercive interventions.

## RESEARCH AND CLINICAL REPORTS

Many of the clinical questions associated with involuntary interventions are amenable to research, and research on these questions has begun recently. Do the treatments provided in conjunction with or pursuant to involuntary intervention have positive clinical outcomes? Are these outcomes associated with reductions in the risk of harm? Are involuntary interventions generally useful, in the long run as well as in the short run, as deterrents to clients harming themselves or others? How does involuntary intervention affect the treatment alliance?

Anecdotally, clients have sometimes expressed to us after-the-fact their appreciation for what they have gained from such external controls as supervised living arrangements that they entered involuntarily or involuntary payeeships. In our experience, an involuntary intervention has sometimes broken the ice of treatment refusal and resistance and has engaged clients with their treatment teams. We have also found that the experience of an involuntary intervention can awaken the client to the physical peril caused by years of substance abuse. Clients in these circumstances report that the experience of an involuntary intervention gave them a powerful message of caring that enhanced their engagement, insight, and personal control and that also engendered their trust in the effectiveness of the treatment team.

Research studies on the subject have begun only recently. The research questions continue to be framed, and more studies are expected (Monahan, Hoge, Lidz, Eisenberg, Bennett, Gardner, Mulvey, & Roth, 1996; Munetz, 1997). The outcomes of involuntary interventions are the subject of a recent study in Manchester, New Hampshire. The authors found clinical improvements for 26 clients who received mandated community treatment under conditional discharge from the hospital (O'Keefe, Potenza, & Mueser, 1997). After one year, clients had improved their medication compliance, housing stability, and vocational activity, and they had reduced their substance use, violent behaviors, and days of hospitalization. At two years, positive effects remained for medication compliance, substance use, and violent behavior, but not for housing stability, vocational activity, and days of hospitalization. Another recent study (Policy Research Associates, Inc., 1998) used a randomized design to evaluate the effect of involuntary outpatient commitment on treatment outcomes for 142 clients being discharged from Bellevue Hospital (NY). All clients received intensive community treatment as well as enhanced assessment and discharge planning. The experimental group received supervision under a court order while controls received the same supervision without a court order. Results showed that the court order itself had no discernible effect on outcomes. Clients had similar outcomes in terms of symptoms, quality of life, treatment continuation, arrests, rehospitalizations, hospital days used, and violent behavior. Regarding violent behavior, the authors note that the study subjects did not include clients considered to be at high risk of violence in the community. Unfortunately, the group sizes were too small to test the significance of an effect related to substance abuse and dependence. Clients with substance use disorders were overrepresented in the court-order group, and this subpopulation apparently had a greater rehospitalization rate.

The underlying question of whether coerced treatment "works" is premature according to Monahan and colleagues, since it is first essential to understand how clients experience coercion (Monahan et al., 1996). In the related field of substance abuse treatment, it has been shown that coerciveness is unrelated to substance abuse treatment outcomes across an array of programs (McLellan, Alterman, Metzger, Grisson, Woody, Luborsky, & O'Brien, 1994). In other words, those who enter treatment involuntarily do as well as those who enter voluntarily.

Research has revealed that perceived coercion may be more important to the treatment alliance than legal coerciveness (Monahan et al., 1995). Individual clients react very differently to coercion, and individuals who are involuntarily admitted to hospitals can in fact feel as if they have not been coerced at all (Lidz, Hoge, Gardner, Bennett, Monahan, Mulvey, & Roth, 1995). More important to the client than the legal status of an involuntary hospital admission may be the admitting staff's attitudes and interactions (Lidz, Mulvey, Arnold, Bennett, & Kirsch, 1993). Clients are more likely to perceive coercion when they believe that they are not mentally ill; when they think that others are not acting in good faith, or are being deceptive, unfair, or disrespectful; or when they feel forced, criminalized, or otherwise poorly treated in the process (Bennett, Lidz, Monahan, Mulvey, Hoge, Roth, & Gardner, 1993; Monahan et al., 1995). In one study, among 105 patients surveyed, 50% felt that forced treatment had generally been in their best interest (Lucksted & Coursey, 1995). Instruments are now being developed to evaluate how the client experiences pressure and force (Lucksted & Coursey, 1995) and how the client perceives control, choice, freedom, and influence in the context of coercive interventions (Gardner, Hoge, Bennett, Roth, Lidz, Monahan, & Mulvey, 1993; Hoge, Lidz, Mulvey, Roth, Bennett, Siminoff, Arnold, & Monahan, 1993; Lidz et al., 1995).

## THE ROLE OF CLOSE MONITORING IN DUAL DISORDERS TREATMENT

In dual disorders services, involuntary interventions fit into a broader array of strategies for close monitoring. *Close monitoring* refers to intensive supervision, which at times is provided with the client's consent and at other times is provided involuntarily. The structures and controls of intensive supervision are so essential a part of dual disorders treatment that we have indicated that close monitoring constitutes a fundamental principle of dual disorders treatment (Drake, Bartels, Teague, Noordsy, & Clark, 1993). Close monitoring and involuntary interventions are more likely to be necessary at earlier stages of dual disorders treatment. As clients gain motivation for treatment and self-control during the later stages, the likelihood of harm diminishes and treatment compliance grows.

We have found that dual disorders treatment is a long-term process that proceeds in four stages (Drake & Noordsy, 1994; McHugo, Drake, Burton, & Ackerson, 1995; Osher and Kofoed, 1989)—engagement, persuasion, active treatment, and relapse prevention. The four stages can be measured (McHugo et al., 1995) and are analogous to the stages of change in addiction treatment proposed by Prochaska, DiClemente, and Norcross, 1992. The engagement stage revolves around developing a trusting relationship with the client. Given that clients' symptoms are least well regulated early in treatment, close monitoring strategies including involuntary interventions in situations of potential harm are more often justified during this stage than during other stages. Monitoring for compliance with psychotropic medications is a common form of close monitoring at this stage, since substance use is not yet a focus of treatment. The persuasion stage focuses on fostering the client's motivation to reduce substance use and to appreciate the adverse consequences of using substances and the positive results of not using substances. During this stage, an important form of close monitoring—urine drug screens—may become helpful. We find that clients often consent to these tests and collaborate in using their findings. The active treatment stage involves helping the client to reduce substance use and to develop active strategies, skills, and relationships for achieving and maintaining abstinence. During this stage, clients learn to use their psychotropic medications voluntarily and begin to reduce their substance use. Again, the likelihood of harm diminishes. Examples of close monitoring during this stage include living in supervised dry housing, urine drug screens, frequent individual sessions and/or home visits, and monitored disulfiram therapy. The relapse prevention stage entails helping the client to develop additional skills and strengths for preventing relapses and coping with setbacks. Close monitoring at this stage may involve regular check-in with a group or an individual clinician and use of a self-help (e.g., AA) sponsor.

Most strategies for close monitoring are voluntary and relatively nonrestrictive. In many cases, the careful use of voluntary close monitoring strategies can prevent progression to involuntary intervention. Clinically, close monitoring provides structure in a person's social environments and supervision over patterns of daily living. Urine and serum drug screens are forms of close monitoring that give clinicians and clients feedback on actual substance use and a basis for the persuasion stage of treatment work, during which the client begins to appreciate the connections between use and consequences. For many individuals with dual disorders, close monitoring enables community-based treatment to progress while self-motivation is still developing. Ideally, the client and the treatment team work together to choose structures and limits and to monitor their effects.

As we will describe, involuntary intervention is morally and clinically justified and necessary in situations where grave danger of harm is present. But otherwise, involuntary interventions are considered only after all voluntary options for close monitoring and engagement in treatment have been exhausted and/or when the client has participated as fully as can be expected in the choice of the intervention. We find, for example, that payeeships can be pursued with the client's active involvement.

## VOLUNTARY OPTIONS

The client's voluntary participation in a carefully designed close monitoring program can in many cases prevent the need for involuntary interventions. Many voluntary options are available for engaging clients in treatment and for supporting treatment adherence. These include counseling (with its many technical variations), treatment plan contracts, informal agreements ("OK, let's agree"), behavioral contracts (such as community reinforcement), and consent to restrictions (see Table 1). Clients can also consent to have legal or programmatic restrictions for a time. Clients can participate in many different ways to show their adherence to the plan. They can keep regular appointments, check in, participate in a drug screen program, enter a supervised living arrangement, and hold to agreements with family members and significant others. Other clinicians and providers, family members, and friends can play important roles as collaborators and supporters for the client's success in the plan.

## RULES OF THUMB

Since many close monitoring mechanisms—and all involuntary interventions—take some responsibility away from the client and place the responsibility outside the person, consideration of such interventions places a special burden on counseling to identify and maximize the client's awareness of their ultimate responsibility for themselves. Clinicians working in New Hampshire have described rules of thumb in close monitoring (see Table 2). For most clients, much time will pass before they develop dependable self-motivation and reliable self-control.

Paying close attention to the client's preferences helps with motivation and self-control and encourages a sense of ownership over the treatment. If the client suggests an alternative that the counselor believes to

TABLE 1. Voluntary Mechanisms for Close Monitoring

**What Types of Voluntary Arrangements Are Made**
- Person-to-person agreement
- Treatment plan
- Behavioral contract
  - Rewards for short-term successes (e.g., praise, going out for a treat)
  - Bonus rewards for longer-term successes (e.g., special time with caregivers or family, new clothes associated with recovery-oriented activities [for a job or for exercise], extra spending money)
- Power of attorney
- Consent for legal protections of property or person
- Consent for more personal supervision (e.g., housing program)

**What is Monitored**
- Drug and alcohol use
- Compliance with taking psychotropic medications
- Psychotropic medication levels
- Attendance at structured activities (e.g., program or competitive employment)
- Level of participation in structured activities
- Avoidance of risky situations

**How Client Participates in Monitoring**
- Meeting in town with provider as scheduled
- Being home to meet with provider as scheduled
- Making scheduled visits to office for appointments
- Participating in laboratory testing for substances or medication levels
- Checking in by phone or in person when having a problem
- Checking in, on a schedule, by phone or in person
- Cooperating in a supervised program (e.g., housing or home-based outreach services)

**Who Collaborates with Client in Monitoring**
- Client's individual substance abuse counselor
- Client's case manager (if different from counselor)
- Treatment team
- Individual team members
- Family members
- Significant others
- Friends
- Other providers (e.g., housing outreach team, vocational specialists, community integration specialist).

be less promising than her own suggestion, the two can agree to try out the client's alternative first, using the model of an empirical trial (see Appendix 1). Expectations must be made clear, measurable, and time-limited. Any restrictive measure, voluntary or involuntary, can and should be time-limited. Restrictions should be shaped for stepwise reductions, beginning as soon as a period of stability has been achieved,

TABLE 2. Rules of Thumb in Close Monitoring

- Do everything possible to preserve trust in the therapeutic relationship
- Maintain unconditional acceptance and positive regard for the client; if blame is necessary, blame the disorder
- Confine the use of involuntary treatment mechanisms to dangerous situations
- Work with the client to try all voluntary options before considering involuntary restrictions
- Consider the risks and the benefits of restrictions, focusing on the consequences for self-motivation and self-control
- Always involve the client in evaluating and choosing the options
- Never surprise the client (e.g., having the police appear at their door to take them to the hospital)
- Be clear on expectations and consequences
- Write a contract
- Limit restrictions as much as possible
  - Provide choices within the restrictive arrangements so the client has some ownership and self-control
  - Limit the amount of restriction as much as possible from the start
  - Plan to reduce the restrictions as soon as possible
  - Impose behaviorally linked time limits on the restrictions
  - Build in steps to freedom, loosening restrictions in stages, so the client can demonstrate capabilities (e.g., after 1 month of full medication compliance, we will reduce daily medication monitoring to every other day)
- Become convinced with the client that the plan is the best option under the circumstances and that the restrictions are required

so that the client gradually takes increased responsibility (time out of hospital, management of money, and the like). Small increments can ensure success and thereby sustain motivation for recovery and self-control. Even when the odds of success for voluntary engagement are low, it is important to try voluntary means. Sometimes the clinician finds that it is necessary to allow the client to experience the failure of his own chosen plans, and that he will subsequently agree to a close monitoring strategy such as a payeeship.

## EVALUATION OF INVOLUNTARY INTERVENTIONS

The principal treatment goals in dual disorders programs include the goals upheld by programs for other clients with severe mental illnesses. Clinicians work with clients to achieve particular treatment effects, such as the control of symptoms or the acquisition of skills, and they also work to preserve safety for the client and others. The principal goals of treatment programs for people with severe mental illnesses also include working with clients to cultivate self-motivation and self-determination and helping clients to develop self-control for recovery. For a client with dual disorders, a proposed involuntary intervention might help to achieve a treatment effect such as the control of symptoms and harmful behavior. The intervention will also likely have an effect—positive or negative—on the client's growing self-motivation and self-control. We suggest evaluating an involuntary intervention through the consideration of three questions. Together, the questions illuminate the clinical indications and contraindications for involuntary intervention with a given client and situation.

1. Is involuntary intervention necessary and likely to prevent harm?
2. How will involuntary intervention affect the client's self-motivation and self-determination?
3. How will involuntary intervention affect the client's self-control?

Each question highlights the impact of involuntary interventions on different goals that operate concurrently in dual disorders treatment. The first question expresses a traditional goal of public mental health systems: *to assure safety*. The second question expresses goals that are typical in psychiatric rehabilitation and CSP services: *to cultivate self-motivation and self-determination for recovery*. The third question expresses a goal that is particularly important in recovery-oriented dual disorders treatment: *to help clients to develop self-control*.

## 1. Is Involuntary Intervention Necessary and Likely to Prevent Harm?

Much of the thinking about involuntary interventions in mental health care follows a conceptual framework in which the goal of assuring safety justifies involuntary restrictions (Group for the Advancement of Psychiatry, 1994). This is consistent with ethical guidance in medical care (Cutler and Gert, 1982). Again, we suggest that the consideration of involuntary intervention related to harm and safety follow the logic of the moral rules: (1) seriousness of harm to be avoided; (2) extent of person's rational responsibility; (3) likelihood of harm; and (4) likelihood that the involuntary intervention will significantly diminish the likelihood of harm.

The clinician and treatment team, with the client whenever possible, should evaluate any imminent risks of harm to the client herself and to others. Exploring voluntary alternatives to the intervention, they may decide that the costs of restricting the client's liberties are less than the costs of the present danger of harm and more generally that the benefits of involuntary intervention outweigh the costs. When the client cannot actively participate in examining the options, the clinician may have to decide with the family (and/or with other members of the client's support network) and with the treatment team whether or not to pursue involuntary intervention. With either choice, the stakes are high. Personal safety and legal liabilities are at stake.

Even those clinicians who most honor individual rights and self-determination are concerned for safety, and they are duty-bound under the legal principles of the *Tarasoff* decision to warn people who are potentially endangered by a client's behavior.[1] Clinicians also have a duty to protect the client's well-being. Having seen the painful results of untreated mental illness or substance abuse—the destroyed relationships, the lost identities, the incarcerations, the increased incidence of HIV infection, the physical degeneration, the deaths—many clinicians see restrictive measures as morally imperative. In this light, one writer has exhorted mental health professionals not to hide behind the "shibboleths" of the dignity of risk, the presumption of competence, or the virtues of choice (Sundram, 1993). Positive treatment for "the casualties of deinstitutionalization," may not be possible without some element of coercion (Mulvey et al., 1987, p. 582). In addition, restrictions of one kind may prevent the necessity to employ even more restrictive measures. A conservatorship, for example, may forestall recourse to involuntary hospitalization. For a very seriously disabled client whose life is degraded, the conservatorship might offer some quality of life or even "transform a dangerous, dysphoric, and deprived existence into one that is relatively free of chaos" (Lamb & Weinberger, 1993, p. 150).

Clinicians must not confuse involuntary interventions with treatment. We encourage clinicians to view an involuntary intervention as a structure that assists in assuring safety during the delivery of definitive treatment for dual disorders. The clinician who must consider a paternalistic act is in the middle—between society and the client. She must stand on this middle ground *in each individual case* and examine how the moral rules apply to the adoption of a paternalistic stance. The involuntary intervention should be kept apart from the treatment and the treatment relationship. The consumer is much more likely to perceive

[1]Details of the *Tarasoff* decision are contained in *People v. Poddar* and in *Tarasoff v. Regents of the University of California*, and are reviewed in Stone (1984). A discussion of the *Tarasoff* decision appears in Applebaum (1994).

the involuntary treatment as motivating if they perceive their clinical team working collaboratively to help them overcome the need for it. The intervention is best placed in the context of the need for people to obey the mandates of the state and the requirements of the society. Treatment, on the other hand, is for the client—something that she needs in order to gain or regain self-control. Our message to the client is, "We are fighting with you against involuntary intervention, and helping you to overcome the need for it."

## 2. How Will Involuntary Intervention Affect the Client's Self-Motivation and Self-Determination?

We then consider the clinical goal of cultivating self-motivation and self-determination for recovery. How will an involuntary intervention affect the client's work on these goals? The lack of treatment motivation occupies a central place in the target problems of clients with dual disorders. The client's motivational state must change for substance abuse treatment goals to be mobilized. Individual substance abuse counseling in dual disorders services therefore focuses centrally on evoking, cultivating, and supporting motivation. The counseling process and motivational development depend on a relationship of trust and empathy between the client and her counselor. In this context, any involuntary intervention could be anathema, and involuntary restriction would seem to undermine the therapeutic endeavor except in situations of grave danger (see Table 3).

Early in dual disorders treatment, though, most clients lack hope, the wellspring of motivation. Clients with severe mental illnesses frequently inhabit a culture of demoralization where social norms sustain disorder and abuse, and where the very idea of hope is foreign. With a client who has little hope, external motivators may be necessary to kindle hope and self-motivation. Medication to control psychiatric symptoms, for example, or a comfortable and well-structured living situation with supports could provide the spark. An involuntary intervention could also give the client the opportunity to experience abstinence, an experience that he can contrast with his otherwise unabated substance use. Nevertheless, the intervention that is involuntarily imposed can damage trust and impair the therapeutic process.

We therefore urge the greatest care in the details of how an involuntary intervention is approached. Indeed, we believe that *how* the involuntary intervention is approached is just as critical as *whether* an involuntary intervention is used. The fullest possible participation of

TABLE 3. Motivational Development vs. Involuntary Interventions

| Motivational development requires … | Involuntary intervention may provoke … | | |
|---|---|---|---|
| Trust in the counselor and in the relationship | Feeling of violation and betrayal<br>Anger at not being trusted | OR | Feeling the counselor cares enough to do something effective |
| Unconditional acceptance by the counselor | Anger over being infantilized and judged when counselor resorts to an authority outside the relationship | OR | Continued nonjudgmental acceptance through the involuntary process with encouragement to grow beyond it |
| Belief in the possibility of change | Sense of failure over efforts already made<br>Loss of self-confidence for making changes<br>Loss of optimism and of the desire for change | OR | Experience of positive change |
| Willingness to try change | Defensiveness and resistance to change | OR | Greater sense of possibility of success |
| Self-efficacy | Perception that others have taken away responsibility<br>Feeling of dependency on caregivers | OR | A sense of control and participation through the involuntary process |

the client is recommended. When an involuntary intervention is carried out such that the client has absolutely no role or control, the intervention may well exacerbate the client's hopelessness and despair and damage the therapeutic alliance. Conversely, when the intervention has been approached with respect and a collaborative stance, we have often seen the therapeutic alliance strengthened and the client's motivation sustained or improved.

Clinicians can often discuss openly with the client the benefits and the risks of using an involuntary intervention. Many clients will see some benefit—staying alive, being cared about (and cared for), being safe without substances; most will see the risks—having responsibility taken away, feeling paranoid about the loss of control, losing trust in the clinician, or wanting to rebel. The probability of the client perceiving the intervention as fair and caring will be increased by encouraging the client's participation as soon as possible. We use a stepwise approach, starting as soon as the possibility of harm appears. After clearly identifying the possible harm and the problems leading to it, we talk specifically with the client about the parameters both for initiating an involuntary intervention and for discontinuing it. We give the client an opportunity to avoid the involuntary intervention by trying to control the dangerous behavior on her own, and we also give intensive training in the skills the client may need to achieve the discontinuation of the intervention (i.e., budgeting skills for a payeeship). When we decide on an involuntary intervention, the clinician and team are very clear and specific about what will happen and why they are doing what they are doing. We keep the dialogue open with the client, make contingency contracts, talk about responsibilities, and plan from the start to build the client's responsibility, skills, and self-control. The involuntary intervention can thus be placed into the context of building self-motivation and self-determination.

## 3. How Will Involuntary Intervention Affect the Client's Self-Control?

Next we consider how an involuntary intervention would affect the client's self-control. Learning to be in control of oneself, including one's treatment and use of substances, is a primary objective for clients in the active treatment and relapse prevention stage. With clients who have dual disorders, clinicians are especially challenged to balance the value of self-directed controls with the value to the client of learning to experience and practice control. Either a major mental disorder or a substance use disorder can cause impaired judgment, disordered behavior, and loss of self-control. One clinical authority has described substance use disorders as a lesion of control, noting that clients with these disorders do not have *reliable* control (Kofoed, 1993). People can consume intoxicants when they have not intended to, or in amounts they did not intend. Like self-motivation, reliable self-control may be foreign to a person with dual disorders. Not knowing self-control, clients with dual disorders are believed to benefit from a temporary experience of external controls, whether they receive these controls voluntarily or involuntarily.

The empathic working alliance and sense of self-determination can be maximized by placing the involuntary intervention not within the clinician's control as a therapeutic tool or punishment, but rather within the legal system as a parameter defined by society that applies to everyone. When these situations occur where society requires this limit be set, the clinician should clearly communicate their willingness to use the therapeutic alliance to help the client overcome the need for the involuntary restriction.

In summary, when the question arises whether some external control should be initiated, we suggest that the clinician consider three questions—how the external structure might assure safety and prevent harm, how it might impair or enhance motivational development and the working alliance, and how it might improve or detract from the client's learning of self-control.

## CASE EXAMPLE

James was a 39-year-old, divorced male. He had a 20-year history of psychosis marked by extreme paranoia, ideas of reference, hallucinations, and delusions of persecution. He had also used alcohol, marijuana, and cocaine heavily in his life, leading to multiple adverse consequences. His diagnoses were Schizophrenia, Chronic Paranoid Type, and Polysubstance Dependence. James was assigned to the assertive community treatment team for dual disorders.

In the spring of 1994, James became quite threatening toward family members and ultimately assaulted one of them, leading to an involuntary hospitalization of two months and a long-term conditional discharge (five years) from the state hospital. The community treatment team initiated the hospitalization and negotiated the criteria for the conditional discharge. During the hospital stay, a guardian was appointed for James. His guardian approved involuntary treatment with antipsychotic medication including, eventually, clozapine and a partial hospital program at the community mental health center following his discharge from the state hospital. These programs gradually resulted in a two-year period of sustained abstinence from substances, freedom from threatening or hostile behavior despite continued delusions, and sustained employment in a supported environment. Close monitoring through medication monitoring, urine drug screens, and frequent visits with the clinical team were gradually tapered off as James demonstrated an independent capacity for self-control and an ability to take personal responsibility for abstinence, medication compliance, and keeping appointments.

As the term of the conditional discharge drew to a close (in the spring of 1999), the team met with James, his guardian, and his family, and came to the mutual decision with James not to apply to the court for an extension of the conditional discharge, but to allow it to expire. Payeeship had earlier been terminated by mutual decision, as James showed he could independently manage his financial resources for his daily living needs.

Over the course of the next six months, James gradually began to drop some of the structures that had supported his stability. He quit his job, became isolated at home, missed occasional doses of medication, and had a few slips with substance use, which showed up on urine screens. The team pointed out this pattern to James and worked with him to prevent a full-blown relapse, to boost his self-motivation, and to maintain his self-determination. However, he became increasingly paranoid toward the treatment team, blamed them for his difficulties, and once again made threats toward family members and other acquaintances. At this point, the team psychiatrist and case manager met with James and advised him that his clinical condition was meeting criteria for an involuntary hospitalization or a renewal of a probate commitment to community treatment. We advised him that we were not in favor of such an action and preferred to develop a voluntary treatment plan that would help him to regain control. Therefore James agreed to temporarily reenter the community mental health center partial hospitalization program, to resume home-based medication monitoring and urine drug screens, and to begin working with a vocational specialist to identify an appropriate vocational structure to provide daily structured activity. By the following morning when he was to start the partial hospital program, James had changed his mind. The treatment team's case manager supported him in following through with the agreement by accompanying him to the partial hospital site and by visiting with him again at lunch hour to give him further support and to remind him of his commitment to regaining stability.

James quickly became acclimated to the program and has stabilized with freedom from threatening behaviors and with clean urine screens. He has also begun to interview for jobs.

## SYNTHESIS

Clients who have dual disorders frequently meet criteria for involuntary intervention. We do not consider involuntary intervention to be part of treatment. We do consider involuntary intervention to be a structure that assists in assuring safety during the delivery of definitive treatment for dual disorders. Involuntary intervention is one strategy for

TABLE 4. Reframing for Self-Motivation and Self-Control

| Black or White | Shades of Grey |
|---|---|
| Either ... Or | Both ... And |
| Directive or Facilitative | Suggesting directions and listening for reflective self-instructions |
| Coercive or Evocative | Providing a structure and asking for further ideas |
| Judgmental or Accepting | Giving opinions and yielding to decisions |
| Critical or Empathetic | Sharing accurate feedback and listening fully |
| Punitive or Permissive | Following through with expected consequences and providing freedom |
| Withholding or Nurturing | Helping to keep valued assets safe and giving freely to help concretely |
| Out of control or In control | Losing control completely in some areas and keeping reliable control in others |
| Consensual or Involuntary | Providing many options to choose from and insisting on a few behaviors |

close monitoring that can be necessary and justified in situations where grave danger of harm is present. We do find it essential to provide different forms of close monitoring through the stages of dual disorders treatment, from engagement, to persuasion, to active treatment, to relapse prevention. Clients will often consent to close monitoring strategies. Empathy, unconditional positive regard, and nurturance are critical for developing motivation, and at the same time disciplines and structures are necessary for developing self-control. In order to address the goals of developing self-motivation and self-control, we offer both liberty and structures in one bundle. With a thoughtful balancing of interventions and shared responsibility, the clinician can assist the client to internalize liberties and nurturance on one side, and discipline and constraints on the other. From clinicians' accounts, back-and-forth dialogue with an individual client leads to a synthesis in which the apparent conflicts between liberty and constraint are reframed and resolved over time (see Table 4).

## CONCLUSION

Involuntary interventions in general should be considered only after exhausting voluntary options for close monitoring and engagement in treatment. Evaluation of a proposed involuntary intervention for a client with dual disorders ought to address several treatment goals simultaneously: the assurance of safety, the encouragement of the client's self-motivation, and the development of the client's self-control. Practice guidelines for involuntary interventions have begun to appear (Torrey & Wysik, 1997). For evidence of those practices that will be most effective, however, virtually no scientific data are available. The approaches suggested here are based on clinical experience with hundreds of clients who have dual disorders. This experience has clarified the wide range of voluntary options that may be used for close monitoring and has indicated some rules of thumb for close monitoring.

## REFERENCES

American Psychiatric Association (1993). *Consent to voluntary hospitalization*. Washington, DC: American Psychiatric Press.

American Psychiatric Association (1987). *Involuntary commitment to outpatient treatment*. Washington, DC: American Psychiatric Press.

Applebaum, P. S. (1994). *Almost a revolution: Mental health law and the limits of change*. New York: Oxford University Press.

Bartels, S. F., Drake, R. E., Wallach, M. A., & Freeman, D.F. (1991). Characteristic hostility in schizophrenic outpatients. *Schizophrenia Bulletin, 17*, 163–171.

Bennett, N. S., Lidz, C. W., Monahan, J., Mulvey, E., Hoge, S. K., Roth, L. H., & Gardner, W. (1993). Inclusion, motivation, and good faith: The morality of coercion in mental hospital admission. *Behavioral Sciences and the Law, 11*, 295–306.

Cuffel, B. J. (1994). Violent and destructive behavior among the severely mentally ill in rural areas: Evidence from Arkansas' community mental health system. *Community Mental Health Journal, 30*, 495–504.

Cuffel, B. J., Shumway, M., & Chouljian, T.L. (1994). A longitudinal study of substance use and community violence in schizophrenia. *Journal of Nervous and Mental Disease, 182*, 704–708.

Culver, C. M., & Gert, B. (1982). *Philosophy in medicine: Conceptual and ethical issues in medicine and psychiatry*. New York: Oxford University Press.

Drake, R. E., Bartels, S. B., Teague, G. B., Noordsy, D. L., & Clark, R. E. (1993). Treatment of substance use disorders in severely mentally ill patients. *Journal of Nervous and Mental Disease, 181*, 606–611.

Drake, R. E., & Noordsy, D. L. (1994). Case management for people with coexisting severe mental disorder and substance use disorder. *Psychiatric Annals, 24*, 427–431.

Gardner, W., Hoge, S. K., Bennett, N., Roth, L. H., Lidz, C. W., Monahan, J., & Mulvey, E. P. (1993). Two scales for measuring patients' perceptions of coercion during mental hospital admission. *Behavioral Sciences and the Law, 11*, 307–321.

Group for the Advancement of Psychiatry (1994). *Forced into treatment: The role of coercion in clinical practice*. Washington, DC: American Psychiatric Press, Inc.

Hoge, S. K., Lidz, C., Mulvey, E., Roth, L., Bennett, N., Siminoff, L., Arnold, R., & Monahan, J. (1993). Patient, family, and staff perceptions of coercion in mental hospital admission: An exploratory study. *Behavioral Sciences and the Law, 11*, 281–293.

Kofoed, L. (1993). Outpatient vs. inpatient treatment for the chronically mentally ill with substance use disorders. *Journal of Addictive Diseases, 12,* 123–137.

Lamb, H. R., & Weinberger, L. E. (1993). Therapeutic use of conservatorship in the treatment of gravely disabled psychiatric patients. *Hospital and Community Psychiatry, 44,* 147–150.

Lidz, C. W., Hoge, S. K., Gardner, W. P., Bennett, N. S., Monahan, J., Mulvey, E. P., & Roth, L. H. (1995). Perceived coercion in mental hospital admission: Pressures and process. *Archives of General Psychiatry, 52,* 1034–1039.

Lidz, C. W., Mulvey, E. P., Arnold, R. P., Bennett, N. S., & Kirsch, B. L. (1993). Coercive interactions in a psychiatric emergency room. *Behavioral Sciences and the Law, 11,* 269–280.

Lindqvist, P., & Allebeck, P. (1994). Schizophrenia and assaultive behaviour: The role of alcohol and drug abuse. *Acta Psychiatrica Scandinavica, 82,* 191–195.

Lucksted, A., & Coursey, R. D. (1995). Therapeutic use of conservatorship in the treatment of gravely disabled psychiatric patients. *Psychiatric Services, 46,* 146–152.

McHugo, G. J., Drake, R. E., Burton, H. L., & Ackerson, T. H. (1995). A scale for assessing the stage of substance abuse treatment in persons with severe mental illness. *Journal of Nervous and Mental Disease, 183,* 762–767.

McLellan, A. T., Alterman, A. I., Metzger, D. S., Grisson, G. R., Woody, G. E., Luborsky, L., & O'Brien, C. P. (1994). Similarity of outcome predictors across opiate, cocaine, and alcohol treatment services. *Journal of Consulting and Clinical Psychology, 62,* 1141–1158.

Monahan, J., Hoge, S. K., Lidz, C. W., Eisenberg, M. M., Bennett, N. S., Gardner, W. P., Mulvey, E. P., & Roth, L. H. (1996). Coercion to inpatient treatment: Initial results and implications for assertive treatment in the community. In D. L. Dennis & J. Monahan (Eds.), *Coercion and aggressive community treatment* (pp. 13–28). New York: Plenum Press.

Monahan, J., Hoge, S. K., Lidz, C., Roth, L. H., Bennett, N., Gardner, W., & Mulvey, E. (1995). Coercion and commitment: Understanding involuntary mental hospital admission. *International Journal of Law and Psychiatry, 18,* 249–263.

Munetz, M. R. (Ed.) (1997). *Can mandatory treatment be therapeutic?* New Directions for Mental Health Services No. 75. San Francisco: Jossey-Bass, Inc.

Mulvey, E. P., Geller, J. L., & Roth L. H. (1987). The promise and peril of involuntary outpatient commitment. *American Psychologist, 42,* 571–584.

O'Keefe, C., Potenza, D. P., & Mueser, K. T. (1997). Treatment outcomes for severely mentally ill patients on conditional discharge to community-based treatment. *Journal of Nervous and Mental Disease, 185,* 409–411.

Osher, F. C., & Kofoed, L. L. (1989). Treatment of patients with psychiatric and psychoactive substance abuse disorders. *Hospital and Community Psychiatry, 40,* 1025–1030.

*People v. Poddar,* 518 P.2d 342 (Cal. 1974).

Policy Research Associates, Inc. (1998). *Final report: Research study of the New York City involuntary outpatient commitment program.* New York: New York City Department of Mental Health, Mental Retardation, and Alcoholism Services.

Prochaska, J. O., DiClemente, C. C., & Norcross, J. C. (1992). In search of how people change: Applications to addictive behaviors. *American Psychologist, 47,* 1102–1114.

Stone, A. A. (1984). *Law, psychiatry and morality.* Washington, DC: American Psychiatric Press, Inc.

Sundram, C. J. (1993). Consumer freedoms and professional responsibility. *Quality of Care No. 56.* (Reprint of the Address of Commission Chairman at the Young Adult Institute Annual Conference, New York, May 6, 1993.)

*Tarasoff v. Regents of the University of California*, 551 P.2d 334 (Cal. 1976).
Torrey, W. C., & Wyzik, P. F. (1997). *New Hampshire practice guidelines for adults in community support programs*. Concord, NH: New Hampshire-Dartmouth Psychiatric Research Center. (Available from PRC, 2 Whipple Place Suite 202, Lebanon, NH 03766.)
Yesavage, J. A., & Zarcone, V. (1983). History of drug abuse and dangerous behavior in inpatient schizophrenics. *Journal of Clinical Psychiatry*, *44*, 259–261.

# APPENDIX 1

## EMPIRICAL TRIAL: EXAMPLE WITH ALTERNATIVE HYPOTHESES

| **CONDITIONS/PROCESS** | Example with Alternative Hypotheses | |
|---|---|---|
| BASELINE DATA | • Client is taking 50% of prescribed doses of medication from Med-minder(TM).<br>• Blood level is subtherapeutic.<br>• Symptoms remain in severe range. | |
| HYPOTHESIS | *Hypothesis1*<br>Client will increase medication adherence by responding to clinicians' reminders and monitoring, and will benefit therapeutically. | *Alternative Hypothesis*<br>Client will increase medication adherence by keeping Med-minder(TM) in the refrigerator, and will benefit therapeutically. |
| OUTCOME INDICATORS | • 90% of Med-minder(TM) doses used<br>• Medication blood level in therapeutic range<br>• Severity of symptoms reduced | |
| DESCRIPTION OF INTERVENTION<br>Duration<br>Activity<br>Frequency<br>Responsibility | *Description for Hypothesis 1*<br>Twice daily for 4 weeks, team members will come to client's home on weekdays, remind him to take medication, and observe him taking it. Client will take medication on his own over the weekends, with family reminders by phone. | *Description for Alternative Hypothesis*<br>For 4 weeks, client will keep Med-minder(TM) in the refrigerator and will adhere to his medication schedule on his own, taking one pill twice a day (as prescribed). Team members will visit client at home as usual and count pills taken. |

| | | |
|---|---|---|
| **MONITORED TRIAL**<br>Quality assurance of intervention<br>Collection of data to indicate outcome | *Monitoring for Hypothesis 1*<br>Team members' log of visits and reminders. Family members' log of reminders to client on weekend.<br>Number of doses missing counted 5 days/week by team members and % taken calculated. Blood level of drug taken by RN during weekly meeting. RN BPRS rating before and after 4 weeks | *Monitoring for Alternative Hypothesis*<br>Number of doses missing counted weekly and % taken calculated at weekly meeting with RN.<br>Random dose count by team during regular home visit. Blood level of drug taken by RN during weekly meeting. RN BPRS rating before and after 4 weeks |
| **[PRELIMINARY FINDINGS]** | Adherence improved to 80%<br>Blood level from subtherapeutic to therapeutic range, BPRS down 5 points | |
| **[FOLLOW-UP]** | *Follow-Up for Hypothesis 1*<br>Reduce visits to 1 × /day, continue plan for 4 more weeks, reduce visits to 3/wk when medication adherence > 90% | *Follow-Up for Alternative Hypothesis*<br>Continue plan for 4 more weeks, or for two weeks after > 90% adherence |

CHAPTER 8

# COERCION IN MENTAL HEALTH CARE

VIRGINIA ALDIGE' HIDAY, MARVIN S. SWARTZ,
JEFFREY W. SWANSON, RANDY BORUM,
AND H. RYAN WAGNER

In the mental health law debate over justification for civil commitment of mentally ill persons, coercion has been a central issue. Participants in the debate have tended to separate into two opposing camps. On one side have been civil rights advocates who argued for minimal deprivations to freedom and choice, and then only in cases of dangerousness and only with formal legal procedures to assure due process (LaFond & Durham, 1992; Morse, 1982; Stefan, 1987). On the other side have been mental health clinicians who wanted to minimize procedural and substantive legal limits on their ability to treat mentally ill persons to alleviate patient symptoms and distress (Miller, 1987; Stone, 1975; Torrey, 1998; Treffert, 1981).

But the lines between the camps are not clear for there have always been a substantial number of mental health clinicians who wanted to avoid invoking the law to compel treatment because of ethical and/or practical reasons. They saw an ethical conflict between their primary function as healers and an enforcer role; and many believed that coercion would preclude development of a therapeutic relationship, consequently interfering with their ability to heal (Meichembaum & Turk,

1987; Miller, 1987; Torrey & Kaplan, 1995). Some of these clinicians have joined their voices with the growing chorus of patient groups which advocate for minimizing any form of coercion in treatment of mentally ill persons (Blanch & Parrish, 1993; Diamond, 1996; Mosher & Burti, 1994). Since adoption of restrictive civil commitment criteria and procedures, and more importantly since imposition of fiscal constraints on psychiatric hospitalization, many other mental health clinicians have turned their attention away from legal processes to noncoercive mechanisms for making treatment possible for mentally ill persons who do not voluntarily seek psychiatric help (see Noordsy, Mercer, & Drake, in this volume.)

Today activist groups, frequently comprising mentally ill persons themselves, have come to the forefront on opposing sides of the coercion issue (Kaufmann, 1999). One side wants to make legal coercion easier for those who need treatment but are unable to recognize their need and to seek treatment voluntarily (NAMI, 1995; Torrey, 1998; Treatment Advocacy Center). The other opposes coercion of any type at all costs, arguing that it violates the basic civil rights of freedom and autonomy, represents an abrogation of the principles of medical care, damages the self esteem of persons with mental illness, and worst, that it is dehumanizing (MadNation, Mancusco, 1997).

## THE NORMATIVE ISSUE

Coercion is both a normative and a practical matter in democracies. It is normative in that democratic societies hold the rights to liberty and self-determination as fundamental principles. Because coercion is antithetical to these basic values, democratic societies allow it only when the state's interests exceed those of the individual. That occurs in two cases: when the state must intervene to protect others (its police power) and when it must intervene to protect the individual (*Parens Patriae* power). Democratic societies, thus, place limits on the state's use of coercion. They allow it only under restricted conditions and only after careful procedures have tried to insure that those conditions are met and that there is 'no other recourse.'

In mental health law, these principles of restricting the state's coercive power were lost to the therapeutic and rehabilitative ideology of the first half of this century (Kittrie, 1971); and it was not until the civil rights movement spread to mental patients in the late 1960's and 1970's that they were restored (Hiday, 1983; LaFond & Durham, 1992). Today civil commitment to a psychiatric treatment facility can occur only when the state gives an individual the basic due process found in other areas of the law, when that individual meets the limited criteria of having both mental illness and dangerousness to self or others, and when there is 'no other recourse.'

'No other recourse' means that there exists no less restrictive alternative to involuntary hospitalization which can obtain the necessary outcomes of protecting the individual and/or others from harm (Chambers, 1972; Hiday & Goodman, 1982). The principle of a less restrictive alternative led to laws on outpatient commitment which mandate psychiatric treatment in the community when the dangerousness of a mentally ill person can be controlled outside an inpatient facility (Hiday & Goodman, 1982; Keilitz, 1990). But since coercion is central to outpatient commitment as well as to involuntary hospitalization, the state is still permitted to impose it only under the conditions of limited criteria (although the criteria may be different than those for involuntary hospitalization), due process, and no other recourse.

## THE PRACTICAL ISSUE

The second issue concerning coercion in democratic societies, the practical issue, raises the question of whether coercion works. Does the recourse of coercion to hospitalization or to community treatment provide the necessary outcomes to justify use of the state's power to take away an individual's freedom: protection of the individual and protection of the public? The practical issue boils down to two questions: Does coercion work as a check to self-harm; and does it work in preventing violence to others?

Some argue that the recourse of coercion in mental health law requires not only prevention of harm but also therapeutic justification, that is treatment of the mental illness (Wexler, 1990; Winick, 1997). The practical question of coercion's working would, thus, need to be expanded to: Does coercion bring about a reduction of symptoms and pain, restore a level of functional capacity, and improve an individual's mental health? And for how long should coercion work?—temporarily while in the hospital or while under mandatory community treatment? Or should recourse to coercion require that there be longer term benefits, reducing the likelihood of harm and the disabilities of mental illness after the time of commitment orders? Beyond these practical legal goals is a broader, humanitarian end: Does coercion work as a means of improving the health and welfare of the individual? Does it lead to an improvement in quality of life? Or, does it undermine achieving these ends?

## LEGAL COERCION: AMOUNT

Involuntary hospitalization is the dominant form of legal coercion in mental health care; consequently it has received the greatest amount of attention and statistical accounting. Although involuntary hospitalization continues to be dominant, fiscal constraints, civil rights laws, community treatment ideology, psychotropic medication, and more recently, managed care have led to dramatic declines in psychiatric inpatient populations over the past five decades. As it has become more difficult to be hospitalized for any reason, many have called for outpatient commitment to compel treatment in the community for persons with severe mental illness (Kress, 2000; Swanson, Swartz, George, Burns, Hiday, Borum, & Wagner, 1997; Swartz, Burns, Hiday, George, Swanson, Wagner, & Landerman, 1995; Torrey, 1998).

The amount of legal coercion to community psychiatric treatment across the nation is uncertain because there is no accurate count. Civil libertarians feared that outpatient commitment would be employed to extend the coercive powers of the state over more individuals than those it could hold in hospitals (Mulvey, Geller, & Roth, 1987; Stefan, 1987). But even though outpatient commitment has been expanded in some states to become a preventive measure for avoiding involuntary hospitalization (while still serving as a less restrictive alternative to involuntary hospitalization), outpatient commitment is not being employed much. We know from a few community and state studies that relatively few persons receive outpatient commitment orders, ranging from zero to 14% of all civil commitment cases (Hiday & Scheid-Cook, 1987; Miller, 1988; Wood & Swanson, 1985). Despite its infrequent use, all but two evaluations found outpatient commitment to have positive outcomes in reducing hospital readmissions, days hospitalized, noncompliance, and in increasing community treatment (Hiday, 1992; Steadman, Gounis, Dennis, Hopper, Rhodes, Swartz, & Robbins, 2001; Swartz et al., 1995; Swartz, Hiday, Swanson, Wagner, Borum, & Burns, 1999).

There are other community coercive legal measures besides outpatient commitment; but information about their use is even more scarce. Conditional release may be the most common form of compulsory community treatment. Physicians frequently use it in discharging a patient from the hospital before the commitment time has lapsed because it allows involuntary rehospitalization with minimal paperwork and without having to begin a new legal procedure (Brackel, Parry, & Weiner, 1985). Some judges use conditional release when they do not want to discharge patients outright (Miller, 1994). It is usually not coercive, in that it amounts to little more than an admonition to take the prescribed medication, go to the mental health center, and stay out of trouble (Hiday, 1995). Revocation generally occurs when family or friends request that a conditionally released patient be returned to the hospital because of symptomatic or dangerous behavior. Sometimes revocation occurs when a patient is returned to the hospital on new commitment papers and it is used to override the new process, thus simplifying readmission. More structure exists in New Hampshire where specific release conditions concerning medication, therapy and substance use are mutually reached by inpatient and outpatient treatment teams and the patient (O'Keefe, Potenza, & Mueser, 1997). In the only outcome study of conditional release, patients on this structured form showed significant and consistent improvement over their prehospital levels in medication compliance, substance abuse and violence for 2 years, and improvement in housing stability, employment and days hospitalized for 1 year (O'Keefe et al., 1997). More structure also exists in conditional release of insanity acquittees, as opposed to civil committees, which involves close monitoring by an assigned agency with quick rehospitalization at the appearance of deterioration or treatment resistance (Bloom, Rogers, Manson, & Williams, 1986; Bloom, Williams, & Bigelow, 1992; Heilbrun & Griffin, 1993; Solomon & Draine, 1995).

Coerced treatment in the community may also occur under legal guardianship, sometimes called conservatorship. When an individual is found to be incompetent, that is, lacking the capacity to make decisions to take care of himself, a court appoints a guardian to make decisions to protect (including care for) that individual. The guardianship may be either full (over all aspects of an individual's life) or partial (over limited parts of his life such as finances or psychiatric care) when incompetence is limited to certain areas. The guardian may determine the need for and require psychiatric treatment on an inpatient or outpatient basis (Diamond, 1996; Geller, McDermeit, Grundzinskas, Lawlor, & Fisher, 1997; Lamb, & Weinberger, 1992; 1993; Young, 1987). A national survey of knowledgeable persons in each state reported that guardianship to coerce psychiatric treatment in the community is not frequently used outside of California (Torrey & Kaplan, 1995). But in one Massachusetts study of mental health center clients, all with severe mental illness, 11% were under a court guardianship order for psychiatric medication and

25% were under full guardianship which could include coerced treatment (Geller, Grundzinskas, McDermeit, Fisher, & Lawlor, 1998). A few studies in California and Massachusetts have found that using guardianship to ensure psychiatric treatment in the community leads to greater stability with fewer hospital admissions and days, and fewer jail incarcerations than control subjects not under guardianship (Geller et al., 1998; Lamb & Weinberger, 1992; 1993).

Some community mental health centers have tried using money management of clients' entitlements to insure treatment and reduce substance abuse. In this form of community coercion the mental health center becomes the representative payee for Supplemental Security Income and Social Security Disability Insurance payments, requiring clients to comply with medication and follow other components of treatment plans in order to receive their money. The center may also make direct payments for housing and other necessities. Many community mental health centers seem to be using this form of outpatient treatment coercion for at least some of their clients while other centers use a representative payee system without making receipt of funds dependent on treatment compliance (Cogswell, 1996; Conrad, Matters, Hanrahan, Luchins, Savage, & Daugherty, 1998; Ries & Dyck, 1997; Rosen & Rosenheck, 1999; Rosenheck, Lam, & Randolph, 1997). Only a few studies have evaluated the effectiveness of these programs; but all report positive outcomes: improved compliance with outpatient treatment (Ries & Comtois, 1997), reduced days hospitalized (Luchins, Hanrahan, Conrad, Savage, Matters, & Shinderman, 1998) and decreased homelessness (Rosenheck et al., 1997).

Although the number of persons ordered to mandatory treatment in the community is uncertain, all studies and state archival data indicate that persons subjected to legal coercion both in and out of hospitals have not changed over the years. Persons who are committed to hospitals or outpatient treatment are overwhelmingly those with few resources in the lower strata of society: the poor, uneducated, unmarried, and members of minority groups (Hiday, 1988; Hiday, Swartz, Swanson, & Wagner, 1997; Hiday, Swartz, Swanson, Borum, & Wagner, 1999; Nicholson, 1986; Nicholson, Ekenstam, & Norwood, 1996) .

## EXTRA LEGAL COERCION

Legal coercion does not tell the whole story of coercion; for there are voluntary patients with no legal constraints who feel coerced and involuntary patients who do not feel coerced. The picture of legally voluntary patients being those with insight into their illness who seek psychiatric hospitalization and involuntary patients being those who lack insight and who must be pushed and dragged into the hospital is a myth.

Legally involuntary patients frequently report that they wanted to be hospitalized at the time of their admission. Some say they would have entered voluntarily if that option had been offered (Monahan, Hoge, Lidz, Eisenberg, Bennett, Gardner, Mulvey, & Roth, 1996). Other legally involuntary patients consciously maneuver to be hospitalized because of difficulty in obtaining voluntary admission and/or because transportation is available only to legally committed patients (Miller, 1982; Monahan et al., 1996). They describe acting on their own by threatening harm or appearing to do something dangerous such as faking suicide, and enlisting the help of others to begin legal commitment procedures. Between 20% and 33% of legally involuntarily hospitalized patients fall into this group of wanting to be admitted at the time of their formal coerced hospitalization (Beck & Golowka, 1988; Edelsohn & Hiday, 1990; Hoyer, 1986; Kane, Quitkin, & Rifkin, 1983; Monahan et al., 1996).

On the opposite side are patients who sign papers saying that they want to be hospitalized but who have been heavily influenced by others in that "voluntary" decision. Many mental health professionals with a distaste for coercion, or who wish to avoid the paperwork and possible court appearances that formal legal coercion may entail, often try to convince reluctant patients to accept voluntary hospitalization (Gilboy & Schmidt, 1971; Reed & Lewis, 1990). Some of their methods, especially threats of involuntary commitment or of incarceration in jail, can be interpreted as coercive. Likewise family, friends and attorneys use threats of commitment, or "no option" coercion to get patients to agree to sign themselves into the hospital (Decker, 1980; 1981; Lewis, Goetz, Shoenfield, Gordon, & Griffin, 1984; Rogers, 1993). Patient surveys indicate that about half of legally voluntary patients report some informal coercion in the process of hospitalization (Beck & Golowka, 1988; Lidz, Mulvey, Arnold, Bennett, & Kirsch, 1993; Rogers, 1993).

Less studied is the occurrence of coercion in outpatient treatment; but the few extant studies report similar findings: significant others encourage, remind and even force persons with severe mental illness to take their medication. Patients describe having a potent awareness of families' and caretakers' responses and sanctions that would follow if they did not take their medication, and being motivated by the knowledge that mental health practitioners could invoke legal measures if they did not follow treatment (Lucksted & Coursey, 1995, Rogers, Day, Williams, Randall, Wood, Healy, & Bentall, 1998).

What is it that comprises extralegal coercion? Pressures that others may exert on mentally ill persons to get them to agree to hospitalization or to comply with treatment in the community can vary from mild persuasive attempts and pleas, through inducements with offers of desired objects or services such as money or a trip, through threats of negative consequences such as involuntary hospitalization or being put out of the house, to strong application of physical force. The MacArthur Coercion studies asked 157 newly admitted voluntary and involuntary patients about these types of pressures and found that about half (46%) reported no pressure of any kind. The largest group who felt pressured reported persuasion (38%) and a mere 4% reported being offered an inducement. Only 19% reported the use of physical force; while another 9% reported threats (Monahan et al., 1996). Where does one draw the line between persuasion and coercion? Can it be drawn objectively or must it be subjectively drawn by each patient? Are there patterns in patients' subjective views of coercion such that an underlying rationale can be discerned?

## THE PERCEPTION OF COERCION

Until recently no researchers attempted to measure coercion. Only in the early 1990s did research measuring coercion appear when the MacArthur Research Network on Mental Health and the Law developed several instruments to quantify patient perceptions of their hospital admission. Two of their instruments, one a fixed choice Admission Experience Survey and one an open-ended Admission Experience Interview, contain items which form two comparable psychometrically sound scales of patients' perceptions of coercion (Gardner, Hoge, Bennett, Roth, Lidz, Monahan, & Mulvey, 1993). These measures essentially define coercion as the opposite of autonomy; thus, feeling coerced in mental hospital admission means perceiving that one does not have influence, control, freedom or choice, or does not make the decision to enter the hospital (Gardner et al., 1993). Such a definition holds coercion distinct from force. Although force is a synonym for coercion in everyday language, it is often used narrowly to mean physical compulsion. The MacArthur definition of coercion, thus, has a broader meaning, reflecting patients' feelings regardless of how they were treated.

Using these scales the MacArthur Research Network reported their major findings: patient perceptions of coercion are not equivalent to legal status; patient accounts of events leading to their hospitalization are as complete and plausible as those of accompanying others and of admission staff but patient perceptions of coercion are slightly different from the other actors; patient perceptions of coercion in the admission process do not change significantly over time; approximately half of patients who did not think they needed hospitalization at admission change their minds to acknowledge such a need in retrospect; negative pressures (threats and physical force) but not positive pressures (persuasion and inducements) produce feelings of coercion; and patient evaluations of the admission process as fair (i.e. others act with impartiality and good faith, and take their views into account) minimize feelings of coercion (Bennett, Lidz, Monahan, Mulvey, Hoge, Roth, & Gardner, 1993; Gardner et al., 1993; Gardner, Lidz, Hoge, Monahan, Eisenberg, Bennett, Mulvey, & Roth, 1999; Hoge, Lidz, Mulvey, Roth, Bennett, Siminoff, Arnold, & Monahan, 1993; Hoge, Lidz, Eisenberg, Gardner, Monahan, Mulvey, Roth, & Bennett, 1997; Lidz et al., 1993; Lidz, Hoge, Gardner, Bennett, Monahan, Mulvey, & Roth, 1995; Lidz, Mulvey, Hoge, Kirsch, Monahan, Bennett, Eisenberg, Gardner, & Roth, 1997; Monahan et al., 1996). Subsequent studies in Oklahoma and New Zealand using the MacArthur Admission Experience Survey and in Florida using their Admission Experience Interview have generally supported these findings (Cascardi & Poythress, 1997; McKenna, Simpson & Laidlaw, 1999; Nicholson et al., 1996).

In our own randomized control trial of outpatient commitment for persons with severe and persistent mental illness, The Duke Mental Health Study ($N = 331$), we were able to ask persons regularly subjected to psychiatric hospitalization (and often to coercion as well) about their perceptions of coercion at two separate times and under two conditions. At baseline, we asked subjects about coercion in their involuntary psychiatric admission to the inpatient facility where we recruited them; and at the end of the study (12 months after discharge into the community from the baseline hospitalization), we asked them about coercion in outpatient treatment (Borum, Swartz, Riley, Swanson, Hiday, & Wagner, 1999; Hiday et al., 1997; Swartz, Swanson, Wagner, Burns, Hiday, & Borum,1999; Swartz, Swanson, Hiday, Wagner, Borum, & Burns, Under Review).

To measure patient perceptions of coercion in hospital admission, we used the perceived coercion items on influence, control, choice, freedom and idea from the MacArthur Admission Survey; and we used a modified version of those items for patient perceptions of coercion in outpatient treatment (see Table 1 for the items of both scales). A simple summation of the five items on each scale yields a score of 0–5, with higher scores representing higher levels of perceived coercion.

Figure 1 presents the distribution of patients' perception of coercion in the hospital admission process. Although all sample members were

TABLE 1. MacArthur Interpersonal Relations Scale

| Original | Modified |
|---|---|
| *Perceived coercion:* | *Perceived coercion:* |
| • I felt free to do what I wanted about coming into the hospital. | • I felt free to do what I wanted about going to the mental health center. |
| • I chose to come into the hospital. | • I chose to go to the mental health center. |
| • It was my idea to come into the hospital. | • It was my idea to go to the mental health center. |
| • I had a lot of control over whether I went into the hospital. | • I had a lot of control over whether I went to the mental health center. |
| • I had more influence than anyone else over whether I came into the hospital. | • I had more influence than anyone else on whether I went to the mental health center. |
| *Negative pressures:* | *Negative pressures:* |
| • People tried to force me to come into the hospital. | • People tried to force me to go to the mental health center. |
| • Someone threatened me to get me to come into the hospital. | • Someone threatened me to get me to go to the mental health center. |
| • Someone physically tried to make me come into the hospital. | • Someone physically tried to get me to go to the mental health center. |
| • I was threatened with commitment. | • I was threatened with commitment. |
| • They said they would make me come into the hospital. | • They said they would make me go to the mental health center. |
| • No one tried to force me to come into the hospital. | • No one tried to force me to go to the mental health center. |
| *Procedural inequity:* | *Procedural inequity:* |
| • I had enough of a chance to say I wanted to come into the hospital. | • I had enough of a chance to say whether I wanted to go to the mental health center. |
| • I got to say what I wanted about coming into the hospital. | • I got to say what I wanted about going to the mental health center. |
| • No one seemed to want to know whether I wanted to come into the hospital. | • No one seemed to want to know whether I wanted to go to the mental health center. |
| • My opinion about coming into the hospital didn't matter. | • My opinion about going to the mental center didn't matter. |

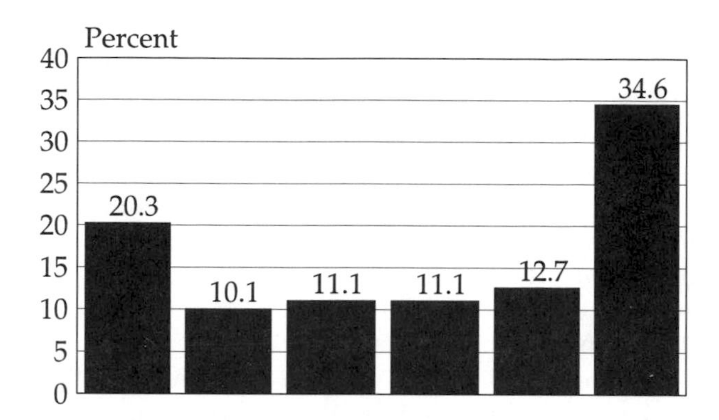


FIGURE 1. Perceived coercion.
Source: Reprinted from (Hiday *et al.*, 1997) with permission from Elsevier Science.

involuntarily admitted, patient perceptions were distributed across the full range of the scale, with over half the responses falling at the two extremes. This bimodal distribution shows just over one-third feeling a high level of coercion and just over one-fifth feeling no coercion, with the rest being equally distributed in the mid-range. The mean of the sample (2.9) is on the high side of the midpoint of the scale.

As the MacArthur studies found, patients' perceptions of coercion in the hospital admission process were highly correlated with their perceptions of threats and physical force, and of fairness, what the MacArthur group calls procedural justice (Lidz et al., 1995). We call our indicator of fairness, process exclusion, because the objective items of the Admission Experience Survey only measure being excluded from the process, that is, they tap not being given a chance to express one's wishes (voice) and not being listened to (validation). (For a full discussion of our sample, the scale and its correlates see Hiday et al., 1997.)

Although all our sample members were involuntarily admitted to the hospital and court ordered to outpatient commitment at discharge, they did not all get outpatient commitment. Rather, just before release, they were randomly assigned either to continue under the outpatient commitment order or to be released from it. Both groups were given case management plus additional treatment according to each patient's individual plan by the local community mental health centers to which patients were ordered for outpatient commitment. Initial orders typically were 45–60 days; but clinicians had the option to seek renewal of the orders from the court; thus, length of time on outpatient commitment could vary up to a full year. Intensity of case management varied as well in number of services and amount of contact. Additionally, subjects could be rehospitalized; and outpatient commitment could be reinitiated. Control group subjects, however, were immunized from outpatient commitment during the year. If inadvertently given an order, control subjects were released from the order. Thus, at the 12 month follow-up those in the outpatient commitment group had varying periods of mandatory treatment in the community; while those in the control group maintained their status of zero days of such treatment.

Figure 2 presents the distributions of control and outpatient committed subjects on the modified perceived coercion scale at the end of the study year. Compared to their perceptions of the hospital admission experience, few patients felt high levels of coercion in outpatient treatment: only 8.4% of the control and 16.7% of the committed subjects felt a high level of coercion. While about twice as many of those in the outpatient commitment group as in the control group expressed feeling a high level of coercion, it was less than half as many of those who felt high coercion in hospital admission. Just over one-fifth of the committed subjects felt no coercion in outpatient treatment, almost the same proportion as in the hospital admission experience; but over half of the controls perceived no coercion. Mean score of the control group (1.3) is significantly lower than that of the outpatient committed group (2.1) (Swartz et al., Under Review). Both of these means of perceived coercion in community treatment are significantly lower than the mean for perceived coercion in the hospital admission process (2.9).

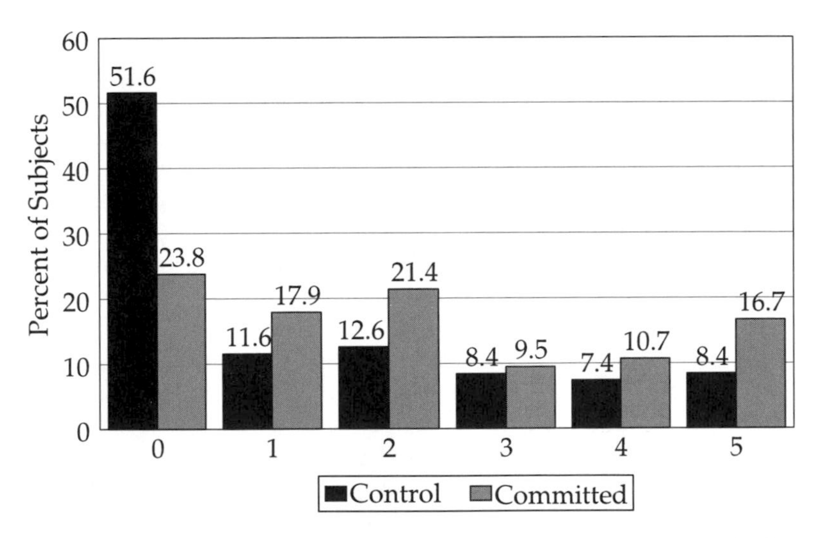


FIGURE 2. Twelve-month perceived coercion scores: controls versus committed subjects.

As in the process of hospital admission, perceived coercion in outpatient treatment is strongly related to the use of negative pressures (threats and force), and to patients' perceptions of fairness (not being validated or not having a chance to be heard). (For a full discussion of this measure of perceived coercion and of a longer, 15 item measure in outpatient treatment, see Swartz et al., 1999.)

Perceived amount of coercion increased as the number of days under outpatient commitment increased (Swartz et al., 1999). Perceived coercion also increased with use of case manger reminders (Swartz et al., 1999). These were statements by case managers about the consequences of treatment nonadherence, such as statements that medication noncompliance and failure to keep mental health center appointments could result in being picked up by the sheriff or in rehospitalization. Sample members might have viewed these reminders as threats to force compliance with treatment, in which case they would be seen as coercive; or they might have viewed them as realistic statements about the consequences of nonadherence. It may also be that case manager reminders are correlated with perceived coercion because these reminders serve as a proxy measure for closer monitoring of subjects on outpatient commitment, and that it is this close monitoring which was perceived by subjects as coercive.

Surprisingly, perceived coercion was not associated with being readmitted to the hospital in the 12 month follow-up or with having legal problems during that time, both of which could be seen as additional coercion in their lives. Perceived coercion was not related to diagnosis or any other clinical variables; but it was significantly related to having insight into one's illness. Expectedly, subjects who had greater insight into their illnesses (as measured by the Insight and Treatment Attitudes Questionnaire, ITAQ, McEvoy, Appelbaum, Apperson, Geller, & Freter, 1989; McEvoy, Apperson, Appelbaum, Ortlip, Brecosky, Hammill, Geller, & Roth, 1989; McEvoy, Freter, Everett, Geller, Appelbaum, Apperson, & Roth, 1989), that is subjects who recognized that they had mental problems and needed treatment for them, had lower levels of perceived coercion; while subjects who had low insight or a view that they were not ill and not in need of treatment reported higher levels of coercion (Swartz et al., 1999).

## SUMMARY AND DISCUSSION

The findings from our data confirm those of previous studies that there is considerable variation in patient perceptions of coercion. This variation is especially notable in our baseline measure of coercion in the hospital admission process, given that sample members were all involuntarily admitted. But it also holds in patient perceptions of coercion in outpatient treatment: both those committed and the controls show variation across the full range of the scale. Such variation reaffirms earlier findings that patient perceptions of coercion are not equivalent to legal coerced status.

Using the formal legal process of civil commitment to hospital treatment or community treatment, thus, will not in and of itself produce feelings of coercion. Rather, the perception of coercion, not merely the legal status of involuntary patient, brings about negative attitudes toward treatment (Rogers, 1993; Shannon, 1976). Without feelings of coercion, formal legal coercive procedures will not necessarily preclude or interfere with a therapeutic alliance as some clinicians fear (Meichenbaum & Turk, 1987; Miller, 1987). Both our and the MacArthur Network's studies indicate that patients perceive high levels of coercion when they are excluded from decision-making and when negative pressures are applied, whether they have either voluntary or involuntary legal status, that is, patients feel coerced when they are not given voice or validation, and when threats and physical force are used in their hospitalization (Hiday et al., 1997; Lidz et al., 1995). The MacArthur Network also found that patients perceive coercion when admission staff and those who bring them to the hospital do not seem to act in good faith and to be concerned about their well being (Bennett et al., 1993). The implications of these findings are clear: to reduce feelings of coercion and to improve the chances of therapeutic results, threats and physical force should be avoided, while concern, respect, fairness, and inclusion should be emphasized.

Our data indicate that perceived coercion is lower in community treatment than in hospital treatment. Although the control subjects who were kept off outpatient commitment during the follow-up had significantly lower levels of perceived coercion than did those in the outpatient commitment group, both groups perceived less coercion in community treatment than they did in the hospital admission process. This finding confirms the common sense basis of outpatient commitment legislation and court decisions that mandatory treatment in the community is a less restrictive alternative to civil commitment to an inpatient psychiatric facility.

The applicable mental health statute in our study does not allow force to be used in giving treatment to persons on outpatient commitment (NCGS § 122C); but the overwhelming majority of our sample believed that outpatient commitment required people to take their medications (82.7%) and keep their appointments (88.6%) (Borum et al., 1999). It could be this belief, reinforced by case manager reminders of the consequences of failure to follow treatment that led to higher levels of perceived coercion among the outpatient committed group than among the controls. Although some may argue that being forced to take psychotropic drugs (which were prescribed for our severely mentally ill sample members) is as coercive as being locked up in a psychiatric facility, our sample judged the coercion experienced in outpatient commitment to be less than that in involuntary hospital admission. By this measure, outpatient commitment, even with accompanying psychotropic medication that patients believe they are required to take, fulfills its role as a less restrictive alternative to involuntary hospitalization.

It should be remembered that outpatient commitment cannot lead to positive outcomes for persons with severe mental illness if the mental health system does not have resources to give the treatment and services that are needed (Swanson et al., 1997). If these persons are to survive at a humane level in the community without relapsing, becoming dangerous and revolving through the court and hospital, they require medication and psychotherapy, the traditional treatments. Also many require and others could benefit from outreach, housing, living skills assistance, day activities, workshops, employment assistance, and crisis intervention. And since patients with severe mental illness who are brought into the civil commitment process tend to be in crisis and have chronic conditions which commonly interact with their environments to produce crises, the treatment and services they receive should be ongoing, designed to be able to avert or minimize crisis situations so that invocation of legal coercion to involuntary hospitalization can be avoided in most cases. In those cases of patients with severe mental illness who do not voluntarily remain in treatment and use needed services, invocation of legal coercion in the form of outpatient commitment may be necessary for extended periods of time; but it should be done with concern and respect to maximize patient inclusion.

## REFERENCES

Beck, J., & Golowka, E., (1988). A study of enforced treatment in relation to Stone's "Thank You" theory. *Behavioral Sciences and the Law, 6*, 559–566.

Bennett, N., Lidz, C., Monahan, J., Mulvey, E., Hoge, S., Roth, L., & Gardner, W. (1993). Inclusion, motivation and good faith: The morality of coercion in mental hospital admission. *Behavioral Sciences and the Law, 11*, 295–306.

Blanch, A., & Parrish, P. (1993). Reports of three roundtable discussions on involuntary interventions. *Psychiatric Rehabilitation and Community Support Monograph, 1*, 1–42.

Bloom, J. D., Rogers, J. L., Manson, S. M., & Williams, M. H. (1986). Lifetime police contacts of discharged psychiatric security review board clients. *International Journal of Law and Psychiatry, 8*, 189–202.

Bloom, J. D., Williams, M. H., & Bigelow, D. A. (1992). The involvement of schizophrenic insanity acquittees in the mental health and criminal justice systems. *Clinical Forensic Psychiatry, 15*, 591–604.

Borum, R., Swartz, M. S., Riley, S., Swanson, J. W., Hiday, V. A., & Wagner, H. R. (1999). Consumer perceptions of involuntary outpatient commitment. *Psychiatric Services, 50*, 1489–1491.

Brakel, J., Parry, J., & Werner, B. (1985). *The Mentally Disabled and the Law* (Ed.). Chicago, American Bar Foundation.

Cascardi, M., & Poythress, N. G. (1997). Correlates of perceived coercion during psychiatric hospital admission. *International Journal of Law and Psychiatry, 20*, 445–458.

Cascardi, M., Poythress, N. G., & Ritterband, I. M. (1998). Stability of psychiatric patients' perception of coercion. *Journal of Clinical Psychology, 53*, 833–840.

Chambers, D. L. (1972). Alternatives to civil commitment of the mentally ill, practical guides and constitutional imperatives. *Michigan Law Review, 70*, 1107–1200.

Cogswell, S. H. (1996). Entitlements, payees and coercion. In D. L. Dennis & J. Monahan (Eds.) *Coercion and aggressive community treatment, A new frontier in mental health law* (pp. 116–171). New York: Plenum Press, pp. 116–171.

Conrad, K. J., Matters, M. D., Hanrahan, P., Luchins, D. J., Savage, C., & Daugherty, B. (1998). Characteristics of persons with mental illness in a representative payee program. *Psychiatric Services, 49*, 1223–1225.

Decker, F. H. (1980). Changes in the legal status of mental patients as waivers of a constitutional right, The problem of consent. *Journal of Psychiatry and Law, 8*, 331–358.

Decker, F. H. (1981). Changes in the legal status of mental patients and hospital management. *Journal of Applied Behavioral Science,* 17, 153–171.

Diamond, R. J. (1996). Coercion and tenacious treatment in the community, Applications to the real world. In D. L. Dennis, & J. Monahan (Eds.) *Coercion and aggressive community treatment, A new frontier in mental health law* (pp. 53–73). New York: Plenum Press.

Edelsohn, G., & Hiday, V. A. (1990). Civil commitment, A range of patient attitudes. *Bulletin of the American Academy of Psychiatry and Law, 18*, 65–77.

Gardner, W., Hoge, S., Bennett, N., Roth, L., Lidz, C., Monahan, J., & Mulvey, E. (1993). Two scales for measuring patients' perception of coercion during mental hospital admission. *Behavioral Sciences and the Law, 11*, 307–321.

Gardner, W., Lidz, C., Hoge, S., Monahan, J., Eisenberg, M., Bennett, N., Mulvey, E., & Roth, L. (1999). Patients' revisions of their beliefs about the need for hospitalization. *American Journal of Psychiatry, 156*, 1385–1391.

Geller, J. L., McDermeit, M., Grudzinskas, J. A., JR., Lawlor, T., & Fisher, W. H. (1997). A competency-based approach to court-ordered outpatient treatment. In M. R. Munetz, (Ed.), *Can mandatory treatment be therapeutic?* San Francisco: Jossey-Bass.

Geller, J. L., Grundzinskas, A. J., McDermeit, M., Fisher, W. H., & Lawlor, T. (1998). The efficacy of involuntary outpatient treatment in Massachusetts. *Administrative Policy in Mental Health, 25*, 271–285.

Gilboy, J., & Schmidt, J. (1971). "Voluntary" hospitalization of the mentally ill. *Northwestern University Law Review, 66*, 429–453.

Heilbrun, K., & Griffin, P. A. (1993). Community based forensic treatment of insanity acquittees. *International Journal of Law and Psychiatry, 16*, 133–150.

Hiday, V. A. (1983). Sociology of mental health law. *Society and Social Research, 67*, 111–128.

Hiday, V. A. (1988). Civil commitment, A review of empirical research. *Behavioral Sciences and the Law, 6*, 15–43.

Hiday, V. A. (1992). Coercion in civil commitment, Process, preferences, and outcome. *International Journal of Law and Psychiatry, 15*, 359–377.

Hiday, V. A. (1995). Outpatient commitment, Official coercion in the community. In D. L. Dennis & J. Monahan (Eds.) *Coercion and aggressive community treatment, A new frontier in mental health law* (pp. 29–47). New York: Plenum Press.

Hiday, V. A., & Goodman, R. R. (1982). The least restrictive alternative to involuntary hospitalization, outpatient commitment, Its use and effectiveness. *Journal of Psychiatry and Law, 10*, 81–96.

Hiday, V. A., & Scheid-Cook, T. L. (1987). The North Carolina experience with outpatient commitment, A critical appraisal. *International Journal of Law and Psychiatry, 10*, 215–232.

Hiday, V. A., Swartz, M. S., Swanson, J. W., & Wagner, H. R. (1997). Patient perceptions of coercion in mental hospital admission. *International Journal of Law and Psychiatry, 20*, 227–241.

Hiday, V. A., Swartz, M. S., Swanson, J. W., Borum, R., & Wagner, H. R. (1999). Criminal victimization of persons with severe mental illness. *Psychiatric Services, 50*, 62–68.

Hoge, S., Lidz, C., Mulvey, E., Roth, L., Bennett, N. Siminoff, A., Arnold, R., & Monahan, J. (1993). Patient, family and staff perceptions of coercion in mental hospital admission, An exploratory study. *Behavioral Sciences and the Law, 11*, 281–293.

Hoge, S., Lidz, C., Eisenberg, M., Gardner, W., Monahan, J., Mulvey, E., Roth, L., & Bennett, N. (1997). Perceptions of coercion in the admission of voluntary and involuntary psychiatric patients. *International Journal of Law and Psychiatry, 20*, 167–181.

Hoyer, G. (1986). Compulsorily admitted patients' ability to make use of their legal rights. *International Journal of Law and Psychiatry, 8*, 413–422.

Kane, J. M., Quitkin, F., & Rifkin, A. (1983). Attitudinal changes of involuntarily committed patients following treatment. *Archives of General Psychiatry, 40*, 374–377.

Kaufmann, C. L. (1999). An introduction to the mental health consumer movement. In A. V. Horwitz, & T. L. Scheid (Eds.) *A handbook for the study of mental health, social contexts, theories, and systems* (pp. 493–507). New York: Cambridge University Press.

Keilitz, I. (1990). Empirical studies of involuntary outpatient civil commitment, Is it working? *Mental and Physical Disability Law Reporter, 14*, 368–379.

Kittrie, N. N. (1971). *The right to be different, deviance and enforced therapy.* Baltimore: The Johns Hopkins Press.

Kress, K. (2000). An argument for assisted outpatient treatment for persons with serious mental illness illustrated with reference to a proposed statute for Iowa. *Iowa Law Review, 85*, 1269–1386.

LaFond, J. Q., & Durham, M. (1992). *Back to asylum.* New York, Oxford University Press.

Lamb, H. R., & Weinberger, L. E. (1992). Conservatorships for gravely disabled psychiatric patients, A 4 year follow-up. *American Journal of Psychiatry, 149*, 909–913.

Lamb, H. R., & Weinberger, L. E. (1993). Therapeutic use of conservatorships in the treatment of gravely disabled psychiatric patients. *Hospital and Community Psychiatry, 44*, 147–150.

Lewis, D. A., Goetz, E., Schoenfield, M., Gordon, A. C., & Griffin, E. (1984). The negotiation of involuntary civil commitment. *Law and Society Review, 18*, 630–649.

Lidz, C., Mulvey, E., Arnold, R., Bennett, N., & Kirsch, B. (1993). Coercive interactions in a psychiatric emergency room. *Behavioral Sciences and the Law, 11*, 269–280.

Lidz, C., Hoge, S., Gardner, W., Bennett, N., Monahan, J., Mulvey E., & Roth, L. (1995). Perceived coercion in mental hospital admission, pressures and process. *Archives of General Psychiatry, 52*, 1034–1039.

Lidz, C., Mulvey, E., Hoge, S., Kirsch, B., Monahan, J., Bennett, N., Eisenberg, M., Gardner, W., & Roth, L. (1997). The validity of mental patients' accounts of coercion-related behaviors in hospital admission process. *Law and Human Behavior, 21*, 361–376.

Lucksted, A., & Coursey, R. (1995). Consumer perceptions of pressure and force in psychiatric treatments. *Psychiatric Services, 46*, 146–152.

Luchins, D. J., Hanrahan, P., Conrad, K. J., Savage, C., Matters, M. D., & Shinderman, M. (1998). An agency based representative payee program and improved community tenure of persons with mental illness. *Psychiatric Services, 49*, 1218–1222.

MadNation. (Website) (1999–2001), www.madnation.org.

Mancuso, L. L. (1997). *Executive summary, involuntary intervention and coerced treatment of people with mental health disorders*. Rockville, MD: US Department of Health and Human Services, Center for Mental Health Services.

McEvoy, J. P., Appelbaum, P. S., Apperson, L. F., Geller, J. L., & Freter, S. (1989). Why must some schizophrenic patients be involuntarily committed? The role of insight. *Comprehensive Psychiatry, 30*, 13–17.

McEvoy, J. P., Apperson, L. F., Appelbaum, P. S., Ortlip, P., Brecosky, J., Hammill, K., Geller, J. L., & Roth, L. (1989). Insight in schizophrenia, Its relationship to acute psychopathology. *Journal of Nervous and Mental Disease, 177*, 43–47.

McEvoy, J. P., Freter, S., Everett, G., Geller, J. L., Appelbaum, P. S., Apperson, L. J., & Roth, L. (1989). Insight and the clinical outcome of schizophrenic patients. *Journal of Nervous and Mental Disease 177*, 48–51.

McKenna, B. G., Simpson, A. I. F., & Laidlaw, T. M. (1999). Patient perceptions of coercion on admission to acute psychiatric services. *International Journal of Law and Psychiatry 22*, 143–153.

Meichenbaum, D., & Turk, D. C. (1987). *Facilitating treatment adherence, A practitioner's guidebook*. New York: Plenum Press.

Miller, R. D. (1982). Voluntary involuntary commitment, The briar-patch syndrome. *Bulletin of the American Academy of Psychiatry and Law, 8*, 305–312.

Miller, R. D. (1987). *Involuntary commitment of the mentally ill in the post-reform era*. Springfield, IL: Charles C. Thomas.

Miller, R. D. 1988. Outpatient civil commitment of the mentally ill, An overview and an update. *Behavioral Sciences and the Law, 6*, 99–118.

Miller, R. D. (1994). Involuntary civil commitment to outpatient treatment. In R. Rosner (Ed.), *Principles and practice of forensic psychiatry* (pp. 118–121). New York: Chapman and Hall.

Monahan, J., Hoge, S., Lidz, C., Eisenberg, M., Bennett, N., Gardner, W., Mulvey, E., & Roth, L. (1996). Coercion to inpatient treatment, initial results and implications for assertive treatment in the community (pp. 13–28). In D. Dennis, & J. Monahan, (Eds.), *Coercion and aggressive community treatment, A new frontier in mental health law*. New York: Plenum Press .

Morse, S. J. (1982). A preference for liberty, The case against involuntary commitment of the mentally disordered. *California Law Review, 70*, 54–106.

Mosher, L., & Burti, L. (1994). *Community mental health, A practical guide*. New York: WW Norton & Co.

Mulvey, E. P., Geller, J. L., & Roth, L. H. (1987). The promise and peril of involuntary outpatient commitment. *American Psychologist, 42*, 571–584.

NAMI (National Alliance for the Mentally Ill). (1995) NAMI statement on involuntary outpatient committment. *American Psychologist, 42*, 571–584.

Nicholson, R. A. (1986). Correlates of commitment status in psychiatric patients. *Psychological Bulletin, 100*, 241–250.

Nicholson, R. A., Ekenstam, C., & Norwood, S. (1996). Coercion and the outcome of psychiatric hospitalization. *International Journal of Law and Psychiatry, 19*, 201–217.

O'Keefe, C., Potenza, D. P., & Mueser, K. T. (1997). Treatment outcomes for severely mentally ill patients on conditional discharge to community based treatment. *Journal of Nervous and Mental Disease, 185*, 409–411.

Reed, S. C., & Lewis, D. A. (1990). The negotiation of voluntary admission in Chicago's state mental hospitals. *Journal of Psychiatry and Law, 18*, 137–163.

Ries, R. K., & Comtois, K. A. (1997). Managing disability benefits as part of treatment for persons with severe mental illness and comorbid drug/alcohol disorders, A comparative study of payee and nonpayee participants. *American Journal on Addictions, 6*, 330–338.

Ries, R. K., & Dyck, D. G. (1997). Representative payee practices of community mental health centers in Washington State. *Psychiatric Services, 48*, 811–814.

Rogers, A. (1993). Coercion and "Voluntary admission", An examination of psychiatric patient views. *Behavioral Sciences and the Law, 11*, 259–268.

Rogers, A., Day, J. C., Williams, B., Randall, F., Wood, P., Healy, D., & Bentall, R.P. (1998). The meaning and management of neuroleptic medication, A study of patients with a diagnosis of schizophrenia. *Social Science and Medicine, 47*, 1313–1323.

Rosen, M. I., & Rosenheck, R. (1999). Substance use and assignment of representative payees. *Psychiatric Services, 50*, 95–98.

Rosenheck, R., Lam, J., & Randolph, F. (1997). Impact of representative payees on substance use by homeless persons with serious mental illness. *Psychiatric Services, 48*, 800–806.

Shannon, P. J. (1976). Coercion and compulsory hospitalization, Some patients' attitudes. *Medical Journal of Australia, 2*, 798–800.

Solomon, P., & Draine, J. (1995). One-year outcomes of a randomized trial of case management with severely mentally ill clients leaving jail. *Evaluation Review, 19*, 256–273.

Steadman, H. J., Gounis, K., Dennis, D., Hopper, K., Roche, B., Swartz, M., & Robbins, P. C. (2001). Assessing the New York City involuntary outpatient commitment pilot program. *Psychiatric Services, 52*, 330–336.

Stephan, S. (1987). Preventive commitment, The concept and its pitfalls. *Mental and Physical Disability Law Reporter, 11*, 288–302.

Stone, A. A. (1975). *Mental health and law, A system in transition.* Rockville, MD: Public Health Service, DHEW Pub. No (ADM), 76–176.

Swanson, J. W., Swartz, M. S., George, L. K., Burns, B. J., Hiday, V. A., Borum, R., & Wagner, H. R. (1997). Interpreting the effectiveness of involuntary outpatient commitment, A conceptual model. *Journal of the American Academy of Psychiatry and Law, 25*, 5–16.

Swartz, M. S., Burns, B. J., Hiday, V. A., George, L. K., Swanson, J. W., Wagner, H. R., & Landerman, R. (1995). New directions in research on involuntary outpatient commitment. *Psychiatric Services, 46*, 381–385.

Swartz, M. S., Hiday, V. A., Swanson, J. W., Wagner, H. R., Borum, R., & Burns, B. J. (1999). Measuring coercion under involuntary outpatient commitment, Initial findings from a randomized controlled trial. *Research in Community and Mental Health, 10*, 57–77.

Swartz, M. S., Swanson, J. W., Wagner, H. R., Burns, B. J., Hiday, V. A., & Borum, R. (1999). Can involuntary outpatient commitment reduce hospital recividism? Findings from a randomized trial with severely mentally ill individuals. *American Journal of Psychiatry, 156*, 1968–1975.

Swartz, M. S., Swanson, J. W., Hiday, V. A., Wagner, H. R., Borum, R., & Burns, B. J. *Under review. The coerciveness of outpatient treatment, Findings among severely mentally ill individuals.*

Torrey, E. F. (1998). *Out of the shadows, Confronting America's mental illness crisis.* New York: John Wiley and Sons.

Torrey, E. F., & Kaplan, R. (1995). A national survey of the use of outpatient commitment. *Psychiatric Services, 46*, 778–784.

Treatment Advocacy Center. (1999–2001). TAC Newsletter. *Tac@po.databack.com*; or www.psychlaws.org. (Website) (1999–2001).

Treffert, D. A. (1981). Dying with their rights on. *American Journal of Psychiatry, 141*, 1041.

Wexler, D. B. (1990). *Therapeutic jurisprudence, The law as a therapeutic agent.* Durham: Carolina Academic Press.

Winick, B. J. (1997). *Therapeutic jurisprudence applied, Essays on mental health law.* Durham: Carolina Academic Press.

Wood, W. K., & Swanson, D. A. L. (1985). Use of outpatient treatment during civil commitment, law and practice in Nebraska. *Journal of Clinical Psychology, 41*, 723–728.

Young, J. L., Mills, M. J., & Sack, R. L. (1987). Civil commitment by conservatorship, the workings of California's law. *Bulletin of the American Academy of Psychiatry and Law, 15*, 127–139.

CHAPTER 9

# WHY ARE SEVERELY MENTALLY ILL PERSONS IN JAIL AND PRISON?

DAVID L. CUTLER, DOUGLAS BIGELOW, VALERIE COLLINS, COURTNEY JACKSON, AND GARY FIELD

*Show me a prison*
*Show me a jail*
*Show me a prison man whose face is growing pale*
*and I'll show you a young man*
*with many reasons why*
*and there but for fortune may go you or I*
—PHIL OCHS (1963)

## PREVALENCE OF SEVERE AND PERSISTENT MENTAL ILLNESSES IN JAILS AND PRISONS

Although the latter half of the 20th century may be viewed as one of the most progressive of the ages for it's development of alternative and less restrictive modes of treatment for the mentally ill. The beginning of the 21st century has been marked by an alarming transinstitutionalization phenomenon of mentally ill people going to jail. This process seems to be occurring simultaneously across the country although in this chapter we will investigate the factors which may account for it in one state, Oregon. Our discussions with key informants, however, suggest that the most powerful factors may in fact be generalizable to other communities.

At the end of 1990s, the news media announced that crime rates had declined nationwide. However, as crime rates decreased during that decade, the prison population simultaneously boomed throughout the country. Steadman et al. (1995) reported that, as of June 1992, there were 444,584 American citizens in jails alone, a 154% increase between 1980 and 1992. According to the Atlantic Monthly (Schlosser, 1998) we now have in this country a "Prison-Industrial Complex" which holds upwards of 1.8 million persons who are poor, homeless, of minority ethnic status (blacks make up about half the prisoners), and mentally ill (including drug dealers and abusers). These people are "raw material" for a rent-a-cell interstate commerce in prisoners. In Oregon, the prison population—and costs—have gone up dramatically over a two decade period in which offenses have not risen, and, furthermore, the prison population of violent offenders has not increased (Whitelaw, 1997).

Teplin (1990) reported that 6.1% of detainees admitted to Cook county jail had a severe mental disorder; among female inmates the proportion was even greater, 12.3%. Among almost two million people currently locked in the nation's penal institutions (jails and prisons) estimates of the prevalence of serious mental illness range from 10% to 25% (Chiles et al., 1990). If these estimates are correct, there are between 200,000 and 500,000 prisoners in the US suffering from major/severe mental illnesses. These numbers are much higher than the 1% or 2% of individuals in the general population thought to have severe and persistent mental illnesses (SPMI). A recent report in Colorado, (Kleinsasser et al.) [which has had a stable state penal system for 12 years (same Governor, Mental Health Director, Chemical Dependency Director)] estimated the percentage of severely mentally ill in the prison population 5 years ago at 6%. The percentage has increased at a rate of about 1% each year to a 1997 figure of 10%.

## Mentally Ill in Local Prisons

Until recently, it has not been easy to count the mentally ill in the prison system in Oregon due to the lack of an effective evaluation mechanism and tracking system. However, in 1996 the mental health department within the corrections system (known as Counseling and Treatment

Services or CTS) began to notice an increase in work load and decided to use a method developed in Colorado for evaluating and tracking severe and persistently mentally ill persons in the prison population. In a recent report to the Oregon E-Board, CTS noted that severely mentally ill persons in the prison population had increased from 13.5% to 15.8% over the last 7 months of the reporting period of 1997 (a 17% increase). By 2001 the number is reportedly over 20%. It is believed these numbers are still increasing. Of those severely mentally ill persons in prison, 40% of the female inmates and 14% of male inmates are currently on psycho tropic medications (the females are mostly on medications for affective disorders).

Oregon Mental Health Division and CTS staff recently compared data bases (Field, 1998). They found that between 1/97 and 6/97, 69% of those newly admitted prisoners meeting criteria for severe mental illness had previously been known to the community mental health system and had received services there. From 7/97 to 12/97 the percentage increased to 85% and from 1/98 to 6/98 it had reached 87%. So it would appear that some individuals, already enrolled in Oregon's community mental health system, have somehow been winding up in the criminal justice system. Furthermore, a careful look at this data reveals that most of this increase is occurring in Multnomah county, Oregon's only large urban county.

Seriously mentally ill persons are also prevalent among the homeless (Goldfinger, 1990). Yet homeless shelters in Portland reported that, during the cold snap of December, 1998, there were more beds available for the homeless than were needed. Usually, when the temperature stays below freezing for several days or weeks, there is a shortage of shelter beds available for all those in need. Where have all the homeless mentally people ill gone? Are they no longer homeless? Have they found other, more secure forms of shelter? Are they in jails and prisons?

There also appears to be an increase in the number of developmentally disabled (DD) persons in the prison population (Field, 1998). Recent studies indicate this is not due to an increase in length of sentence, but to an increased number of people being sentenced (Petersilia, 1997). It is possible that IQ's of these persons may be too high to qualify for treatment within the "DD" system of care and support in community programs, but also too low for them to cope in a competitive community environment on their own. Some of these developmentally disabled inmates also have co-occurring mental illnesses, although they are not classifiable as persons with severe mental illnesses.

## MENTAL HEALTH SYSTEM FACTORS

What is driving this problematic migration of mentally ill persons into the justice system? One factor influencing these numbers in prisons could be changing numbers of individuals going through all the various stages of the commitment process. It has always been difficult to get committed in Multnomah county (Oregon's largest urban county) and perhaps this difficulty has resulted in fewer violent mentally ill being committed, leaving no other choice for local officials but incarceration.

Another factor could be lack of support. There has been considerable research that shows that mentally ill patients have smaller support networks, (Cutler et al., 1987; Pattison et al., 1979; Tolsdorf, 1978), require daily structure and support (Strauss et al., 1985), are highly stress sensitive (Ganzini et al., 1990; Masuda et al., 1978), frequently stop taking medicine for a variety of reasons (Stein and Test 1980), and require appropriate housing (Cutler, 1986). Furthermore, a substantial proportion of SPMI persons need housing and other supports for very long, if not life long periods (Bigelow, 1998; Trieman et al., 1998 *CMHJ 34*(4) 403, 405, 407, 417). Without these needs being met for them systematically, they are prone to get in trouble (Cirincione et al., 1994).

Masuda et al. (1978) reported that high levels of life stress leads to incarceration. Severely mentally ill persons are considerably more stress sensitive than the general population. As well, they are at greater risk of being stressed by unstable housing, social, health and working circumstances. They need support in the community at least as much as they do freedom from restrictive state hospital care, but they don't always get support, even when trained professional provide the service (Biegel et al., 1994). This increasing exposure to stress in community living may be a factor driving SPMI people into the justice system.

In an article published in 1992 (Cutler et al.) we predicted that fragmented funding streams could destroy the support systems for highly vulnerable severely mentally ill people and unfortunately those predictions have been disturbingly accurate. Current "managed care" funding schemes under the Oregon Health Plan which "incentivize" brief or less intensive treatment over intensive and indefinite care may be driving long term mentally ill people to "bad behavior" in order to be cared for in some fashion. It is possible that erosion of accessible intensive community support services is causing a "transinstitutionalization" to the correctional system (Cutler et al., 1998). This is in spite of the fact that our own study (Clarke, Herrinkx, Kinney, Paulson,

Cutler, Lewis, & Oxman, 2000) suggested that assertive community treatment teams can reduce both hospital utilization and incarceration.

Field (1998) noted that, without continuity between prison treatment programs and community treatment, especially drug abuse oriented, mentally ill offenders are likely to re-offend. Most community mental health services do not tailor programs to mentally ill offenders. Housing programs, in particular, are likely to try to avoid them altogether, not wanting to mix former inmates with other vulnerable mentally ill persons. Therefore, once a severely mentally ill person has become involved in the criminal justice system it becomes even less likely that their needs for support and treatment will be met in the mental health system. Some examples do exist of crisis systems designed to connect with jails in order to divert mentally ill to more appropriate treatment (Lambert, 1999; Leopold, 2000) but many more are needed and diversion is also dependent on the availability of alternative beds.

## AVAILABILITY OF MENTAL HEALTH BEDS

The rising population of mentally ill prisoners coincides with reduction (as of 1998) of state hospital non-forensic beds in Oregon to the lowest level since 1877. Back in 1877 the first asylum, The Hawthorne Hospital in Portland, was caring for 230 patients at a cost (to the state

TABLE 1. Service Penetration: General Population and Estimated Persons with Serious and Persistent Mentally Illnesses (2% of GP)

| Year | Population | | State hospitals | | | Community mental health programs | | |
|---|---|---|---|---|---|---|---|---|
| | General population | SPMI persons | ADP | As a of % GP | As a % of SPMI persons | Caseload | As a % of GP | As a % of SPMI persons |
| 1959 | 1,777,000 | 35,540 | 4,900 | 0.28 | 13.79 | 30,640 | 1.72 | 86.21 |
| 1979 | 2,544,000 | 50,880 | 1,100 | 0.04 | 2.16 | 49,780 | 1.96 | 97.84 |
| 1988 | 2,690,000 | 53,800 | 1,000 | 0.04 | 1.86 | 52,800 | 1.96 | 98.14 |
| 1995 | 2,842,321 | 56,846 | 788 | 0.03 | 1.39 | 56,058 | 1.97 | 98.61 |

Note: GP is general population of Oregon; SPMI are estimated number of serious and persistent mentally ill persons; ADP is average daily population of state hospitals; CMHP are community mental health programs.

TABLE 2. Intensive Mental Health Inpatient Beds in Oregon

| Type of facility/bed | 1988 | 1998 |
|---|---|---|
| State hospital | 515 | 198 |
| Local acute care | 8 | 140 |
| Crisis respite | 16 | 75 |
| Community extended care | 0 | 391* |
| TOTAL | 539 | 804 |

Note: From Mental Health and Developmentally Disabled Services Division, January 4, 1999.
* Even though there are many more of these beds than there are state hospital beds they do not turn over and therefore do not serve as many individuals as the hospital beds did. These beds are generally paid for in a variety of ways which can capture Medicaid and other federal funding streams rather than the 100% state funding enjoyed by state hospital beds.

of Oregon's 100,000 inhabitants) of $5.50/person/week, at that time 52% of the state budget (Larsell, 1945)! Currently, there are less than 200 state hospital beds for general adults (not counting child, geriatric or forensic state hospital beds) in a state with over 3 million people.

The closure of Dammasch State Hospital, and the downsizing of the remaining state hospitals as well as the Fairview Training Center, have affected the residential opportunities for SPMI persons and Developmentally Disabled populations. State hospital beds have been reduced, while other sorts of beds have been added in the community tailored to the needs of those patients who were in long stay slots in 1995 when Dammasch closed. Table 2 shows 1998 beds at 149% of the 1988 level. In accord with the plan to migrate from state to local institutions, state institutional beds (acute and long term) are at 65% (338 beds) of the 1988 level. Crisis beds are at 469% (75 beds). Most important, a new category of intensive long stay bed, Community Extended Care, are at 391 beds. Extended, State and Acute, in combination, are at 142% (729) of 1988 levels. Therefore, Oregon should be better able to serve these folks with intensive residential beds in 1998 than it was in 1995. But, is this improved resource adequate to prevent caseload shifting to the correctional system?

## HOW MANY BEDS REALLY ARE NEEDED?

The Canadian province of British Columbia (BC) is very similar in population to Oregon and has about the same outpatient caseload (Bigelow, Sladen-Dew, & Russell, 1994). But BC had about as many

TABLE 3. Mental Health Services Beds in British Columbia

| Type of facility/bed | 1987/88 | 1996/97 |
|---|---|---|
| Residential care | 1,733 | 1,755 |
| Family care | 133 | 158 |
| Supported housing | 279 | 2,352* |
| Emergency shelter | 70 | 132** |
| Sub total (Non-intensive) | 2,215 | 4,397 |
| Acute inpatient | 725 | 717 |
| Crisis residential | 10 | 83 |
| Provincial hospital | 1,220 | 808 |
| Tertiary | 550 | 666*** |
| Subtotal (Intensive) | 2,505 | 2,274 |
| TOTAL | 4,720 | 6,671 |

*an estimated 2,600 more are needed.
**an estimated 35 more are needed.
***most additional tertiary beds in planning stage.
From: Adult Mental Health Division, Ministry of Health and Ministry Responsible for Seniors, 1998.

beds in the provincial hospital as Oregon has in hospitals and community intensive inpatient facilities combined (Tables 2 and 3). BC also has as many acute psychiatric beds (in local general hospitals) as Oregon has intensive inpatient beds in total. In addition, BC has about twice as many tertiary beds as Oregon has hospital beds. This brings the total of intensive beds to about three times as many. BC also has several times the number of less intensive residential beds as Oregon. Although BC is under-resourced, itself, it is less under-resourced than Oregon and comes closer to meeting the needs of its severely ill population (Torrey, Bigelow, & Sladen-Dew, 1993). The BC comparison suggests that, even with recent improvements in Oregon, there is a dearth of beds and that lack of mental health beds may be a significant contributor to diversion of mentally ill persons into justice system beds.

## THE UNIQUE AND UBIQUITOUS ROLE OF STATE HOSPITAL BEDS

State hospital and other acute care beds exist ostensibly for treatment purposes. Although, the migration from state hospitals to other beds appears to have enriched the community service system considerably, and even though there are many more beds of various sorts now, these newer beds lack certain unplanned for ubiquitous characteristics which the old state hospital pocessed. In some fashion not entirely clear, the state hospital beds, despite their relatively small numbers functioned as a flexible "last resort" capability with sufficient holding capacity to quietly remove persons from unmanageable situations. Alas, there is no, or much less, such magical capacity in the new system.

As the number of state hospital beds fell, the admissions to local hospitals increased dramatically. These alternative hospital beds were used for brief treatment. Very few of these new beds are able to provide 24 hour daily structure and support needed by many SPMI people on a long term basis (Bigelow & Cutler et al., 1987). Although the state has purchased a number of these beds for short stays, for many disabled persons, they serve only to temporarily juggle them and then must discharge them as quickly as possible or local hospitals lose money.

## HOUSING CAPACITY

Baker and Douglas (1990) and Leff et al. (1996) reported that the appropriateness of housing to the needs of severely mentally ill persons had a lot to do with their likelihood of successful adaptation to the community and even the effectiveness of psychiatric medication.

In a recent study of 961 severely mentally ill individuals admitted to three types of supported housing utilized in Oregon, Skryha et al. (1999) found that 36.5% of the clients in the study had previous criminal justice involvement, but after living in one or another form of supported housing for an average of 3+ years their involvement dropped to 5.6%. Most of these same individuals, in the five years prior to finding housing, had been homeless or changed residences frequently. Such a lack of stability is a likely substrate for minor criminal behavior such as vagrancy, trespassing, and shoplifting. These activities might lead to other behaviors such as resisting arrest, or menacing or assault which could result in serious enough charges to insure incarceration.

## CHARACTERISTICS OF MENTALLY ILL PERSONS IN JAILS AND PRISONS

Mentally disabled persons like the rest of us are especially in need of money, housing, and friends. Much of their behavior can be understood as (1) a struggle to meet these needs and (2) a struggle in the opportunity structure in which they find themselves (cf., Jessor et al.) On their own, income is often minimal and intermittent. This places them in the proximity of criminal enterprise and at risk of unlawful attempts to acquire money and other subsistence. In the absence of money, even acquiring food can become a struggle. Some homeless mentally ill persons are known to subsist entirely from garbage disposals behind fast food outlets.

In a study conducted in 1985 and reported in Community Mental Health Journal Belcher (1988) reported that 64% of patients discharged from a state hospital in a Midwestern city of 1.5 million, and who had a history both of chronic mental illness and homelessness, became involved with the criminal justice system soon after discharge. These people were also observed to have poor job skills and poor social supports.

Minkoff (1987) and Lamb (1981) predicted that deinstitutionalization would have profound effects on the character and behavior of SPMI persons, especially those who never were institutionalized. In the absence of adequate supervision and wrap-around support, the effect may have been even more profound than thought.

Epidemiological surveys reveal the existence of many SPMI persons who never come to the attention of mental health or other services (Robins & Regier, 1991). Changing social conditions may have caused these these otherwise invisible SPMI persons to drift into the criminal justice system. Social changes which may be affecting epidemiological rates include, for example, an increasing incidence of MRDD, Borderline PD, and/or Depression in the general population as a result of unwanted pregnancies, neonatal alcohol syndrome, child abuse, and violence in the family in general. SPMI people are more likely to be caught up in these social pathologies, and more deeply affected by them. Even though crime rates are down, the perception of violence in the culture may have led to an increase in incarcerations for less serious offenses, drawing these people into the criminal world. The drift into the criminal justice system is a long-term, profound socialization process. We should not be surprised some to find mentally ill persons pretty thoroughly criminalized after a decade or two of exposure.

Torrey reports (Treatment Advocacy Center, 3/18/99, http://www.psychlaws.org) that about 40% of 3.5 million Americans who suffer from schizophrenia and manic-depressive illness are not receiving treatment on any given day, resulting in increased rates of violence, homelessness, and suicide. These individuals commit nearly 1,000 homicides per year or about five times the per capita average. About 200,000 homeless persons have these mental illnesses and 28% of them forage for some of their food in garbage cans. Torrey estimates jail and prison costs for these SPMI people at a minimum of $8.5 billion per year.

While substance abuse has become notorious, it has also become accepted and accessible. Mental and emotional comfort (symptom relief) is high on the priority list. This inclines some to use whatever substances they think may reduce the discomfort, induce a little joy, and contribute to group acceptance. Consequently, many persons with severe mental illnesses also have co-occurring substance abuse problems. The literature suggests in fact that substance abuse has more to do with violent and criminal behavior than mental illness (Klassen & O'Connor, 1988; Reiger et al., 1990). Mentally ill people are more likely to be detrimentally affected by any psycho-active substance they might use, and this could cause them to run a foul of the law. Munetz (2001), in a recent study reported that 70% of severely mentally disabled prisoners were actively abusing substances at the time of incarceration.

These conditions marginalize people and render them liable to attention by the justice system. The principal means of preventing this fate for these people is to provide them with alternative homes, income, and social environments. For a significant number, secure environments are needed (Lamb, 1980; Munetz, Peterson, & Vandershie, 1996). There are treatment modalities proven to improve the lives of individuals suffering from these persistent mental disorders and related disabilities (Stein et al., 1981; Biegel et al., 1994; Clarke et al., 2000), but for one reason or another we seem to be failing to provide enough of that treatment to those who need it in order to function lawfully and adequately in the community. In the absence of available treatment for mental illness, with or without coexisting substance abuse, far too many individuals end up in jail or prison rather than in appropriate treatment. This problem is certainly not helped by the effects of poverty and the de-emphasis on housing and social welfare programs that previously provided a broader "safety net" for vulnerable individuals.

## A POSSIBLE ALTERNATIVE? THE PSYCHIATRIC SECURITY REVIEW BOARD (PSRB)

For the past 25 years Oregon has had a model program for the management of mentally ill offenders known as the Psychiatric Security Review Board (PSRB) (Rogers et al., 1986). The program was instituted in the mid 70's in response to public concern that the old system placed public safety in jeopardy as a result of premature release and lack of community monitoring of conditionally released {civilly committed?} adjudicated insanity acquittees.

The PSRB is composed of a five member part-time board consisting of a lawyer, a psychiatrist, psychologist, a parole/probation professional, and a lay person all appointed by the governor to serve four year terms. The PSRB is independent of both the court system and the mental health system, although the Mental Health and Developmental Disabilities Services Division (MHDDSD) is responsible to provide or fund mental health services to this population, both in the state hospital forensic units and in community programs. Following an insanity verdict (Guilty Except For Insanity) (GEFI) the judge determines whether the person shall be placed under the jurisdiction of the PSRB and, if so, whether the placement will begin in the forensic hospital or on conditional release. The length of the sentence is determined by the maximum sentence the person could have received if found guilty of the specific crimes charged. (Note that this sentence is much longer than the likely time in prison, had the person been routed to the corrections system.)

Once assigned to PSRB, the assignee's placements and movements through the hospital and the community are under the control of the PSRB. The PSRB may discharge an individual prior to the end of the insanity sentence, if the individual is deemed no longer mentally ill nor a danger. There is a caveat: persons with SPMI are deemed in the legislation to be potentially mentally ill even in symptom remission, and, therefore, not eligible for early discharge. About two thirds of the PSRB population have schizophrenia.

The program has proven to be effective (Bloom & Bloom, 1981; Bloom et al., 1986, 1991; McFarland & Bigelow, 1993) for those it manages. Consequently, the PSRB program might seem a suitable alternative for mentally ill persons incarcerated in the correctional system. But there are barriers to this otherwise attractive alternative. Defendants in court are often advised by their lawyers against the insanity defense, due to the potential greater length of the sentence. Furthermore, expanding the PSRB program to include more of people currently routed to the corrections system may require change in the law governing PSRB and the availability of much greater funding for forensic hospital beds.

## CHARACTERISTICS OF THE JUDICIAL SYSTEM AS AN OPTION FOR SPMI PEOPLE

Jails and prisons are a kind of "safety net"—home and family, emergency shelter and care. Some persons who have not succeeded in the community—for want of adequate mental health care or whatever reasons—find there a building which provides shelter and meals, a group of people which includes friends, and a certain amount of care. In this sense, jails and prisons serve as an alternative support and treatment system for SPMI.

Another set of possible causes relates to various changes in the criminal justice system. For example: Community policing and an increase of officers may have led to an increase in arrests and incarceration resulting from behaviors arising from mental illness symptoms. Police in Portland have a greater understanding of mental illness due to the Crisis Intervention Team (CIT) training program, but it continues to be easier for police to put people in jail than the hospital which requires a substantial level of symptomotology to trigger an admission.

The Criminal Justice System, consistent with popular sentiment, has become increasingly strict and less discriminating. Mental illness is less likely to result in alternative dispositions at any stage of the process, beginning with the investigation. (In previous times, the police would have more likely taken the apparently SPMI person to a mental health service and dropped the matter.)

## HYPOTHESES—WHY THE INCREASE IN MENTALLY ILL INMATES?

Thus we have three places to look for causes of the high prevalence of SPMI persons in jails and prisons. We hypothesize that the following variables increase the likelihood of a SPMI person being in jail or prison.

Characteristics of the mental health system:

lack of sufficient, appropriate housing
lack of sufficient support
lack of sufficient treatment, medications, crisis services, and hospital beds

Characteristics of the criminal justice system:

restricted commitment
more accessible support and care
easier access for police officers
strict sentencing
lack of effective aftercare

Individual characteristics:

unmet needs for housing, resources, and friendships
substance abuse
criminal history
criminal attitudes
criminal behavior

## METHODOLOGY

Although these hypotheses require a major research program involving a complex series of studies, to begin with, we decided to ask key participant-observers in Oregons only metropolitan center to help us confirm plausible explanations for the migration of seriously and persistently mentally ill people into jails and prisons. We were able to hire a Stanley foundation scholar medical student (CJ) and recruited a senior psychiatry resident (VC) during the summer of 1999 to assist in the developing and conducting a structured interview for our conversations with participant-observers. We then contacted key informants in a variety of settings in the metropolitan mental health service system, including the state prison, the forensic unit of the state hospital, probation and parole, the county jail and the county mental health program office. We conducted our structured conversations with 26 persons some of whom were consumers, psychiatrists, case managers and discharge planners, as well as jail and prison staff.

## FINDINGS

There was a substantial amount of similarity in the views expressed by these various observers from quite different interests and locations in the service system. The single most highly scored reason for incarceration was **lack of affordable housing** for mentally ill people, followed very closely by a **lack of a personal support systems.** Both averaged 4.3 out of a possible score of 5. According to a local highly respected program manager;

> There is a lack of specialized housing in Multnomah Co, there might not be a decrease, but a good study might show a decrease in housing relative to the increase in population of clients.

The third highest item was **drug use/abuse** at 4.1. According to a consumer who had been in prison

> Drug use is the "number one factor for ending up in jail, including mental illnesses. It's very difficult to quit drug abuse when you have a mental illness."

Right behind ranked 4th and 5th, at a score of 4.0, were **reduced availability of services** from mental health agencies *"There is no aggresive presence to keep clients under the care of MH Services."* and **lack of medication compliance**. Next tied at 3.8 was a **lack of coordination** among agencies (*Many clients have had contact with MHS but are out of contact and have been for quite a while.*) and the **national trend towards incarceration**. At a score of 3.6 was the **lack of long term and short term beds** *"There is a tremendous shortage of beds for people, who are difficult to manage"*. All of the other items scored below 3.5 which probably means that although they may be factors they may not be as important.

## CONCLUSIONS

Although the number of participant observers we were able to interview was limited, they were all strategically located and each had important stakeholder interests as well as first hand experience with severely mentally ill persons of this community. In their collective view, housing and lack of supports for living are major factors driving migration into the criminal justice system. This appears to be seen as more important than the availability and coordination of formal mental health services including non compliance with prescribed medication. Substance abuse is also a major compounding factor. Notably, justice system factors and lack of hospital beds are further down the list of factors. This collective view is consistent with the Community Support Program notion that successful community living is dependent upon adequate housing and other supports. It is also quite consistent with what we have learned from the liturature.

Based on the views of these participant observers, we suspect that there are a number of obvious and ordinary things that one could imagine might help to stem the tide of SPMI people into the correctional system. First on our list would be to develop a greater housing capacity across the range of options from supported living to secure facilities. To be effective this housing resource also needs to cope with or have close ties to programs which can deal with substance use/abuse without simply ejecting people and will need to be accessable to mobile crisis teams and other jail diversion projects 24 hours a day seven days a week just as the jails are. The development of personal support systems which included meaningful daily activities such as "individual placement and support" back to work programs could also be implicitly included in the design of or in association with such housing. Better funding for dual diagnosis programs would be important if these sorts of clients are going to have a chance to hold on to their housing and their work. Efforts also could be made to help shore up the struggling community agencies so that they have enough energy to provide the necessary efforts needed to assure collaboration and coordination across agency boundaries for complicated and difficult to treat high utilizer sorts of clients. Finally, some new transitional housing needs to be invented to fill the need (formerly met by those ubiquitous but now scarce state hospital beds) which is not well met by current inadequate crisis, transitional, and longer term beds. The resources outlined above could conceivably be designed to be cheaper, on a per person basis, than either hospital or criminal justice alternatives.

The status quo option will no doubt continue in the short run because it has the advantage of shifting the costs, services, and the SPMI people, themselves, into the correctional system and its jails and prisons, as a disposition of last resort until they too begin to complain more vociferously. The correctional system despite increasing it's capacity to provide in house mental health services offers little or no prospect of rehabilitating or even ameliorating mental illnesses. Worse, this shift of SPMI people into the criminal subculture compounds unremediated mental illnesses with criminalizing influences. That, together, with many other problems; financial, physical illness, homelessness, could in the long run create serious pathology for Americas cities to say nothing of the moral failure that it demonstrates as a society.

It is now established that persons with severe mental illnesses are flowing into the criminal justice system and thereby into its jails and prisons. Further definition of that migration is essential to the health of our society as well as that of the mentally ill persons, themselves. We need to know how that migration works, what the drivers are, and what we can do about it. This project gathered data from those who have the clearest views of the problem. Their reports may help to see reasonable alternatives which perhaps may not be new but which with substantial renewed effort should help to cut off this flow into the jails and prisons and establish support systems in the community for desperate mentally ill persons.

## REFERENCES

Adult Mental Health Division, Ministry of Health and Ministry Responsible for Seniors (1998). *Revitalizing and rebalancing British Columbia's mental health system*. Victoria, BC: Queen's Printer for BC. 53.

Baker, F., & Douglas C. (1990). Housing Environments and Community Adjustment of Severely Mentally Ill Persons. *Community Mental Health Journal, 26*(6), 497–505.

Belcher, J. R. (1998). Are the Jails Replacing the Mental health System for the Homeless Mentally Ill? *Community Mental Health Journal, 24*(3), 185–195.

Biegel, D. E., Tracy, E. M., & Corvo, K. (1994). Strengthening Social Networks: Intervention strategies for mental health case managers. *Health and Social Work, 19*(3), 207–216.

Bigelow, D. A., Cutler D. L., Moore, L. J., McComb, P., & Leung, P. (1988). Characteristics of hard-to-place patients in a state hospital. *Hospital & Community Psychiatry*.

Bigelow, D., Sladen-Dew, N., & Russell, J. (Spring, 1994). Serving severely mentally ill people in a major Canadian city. In Lamb, R. (Series Ed.), Bachrach, L., Goering, P., & Wasylenki, D. (Vol. Eds.). *Mental health care in Canada'*. No. 61. New Directions for Mental Health Services. 53–62. San Francisco: Jossey-Bass.

Bloom, J., & Bloom, J. L. (1981). Disposition of insanity defense cases in Oregon. *American Academy of Psychiatry & Law Bulletin*, 9, 93–99.

Bloom, J., Williams, M., Rogers, J., et al. (1986). Evaluation and treatment of insanity acquitees in the community. *Bulletin of the American Academy of Psychiatry & Law, 14*, 231–244.

Bloom, J., Williams, M., & Bigelow, D. (1991). Monitored conditional release of persons found not guilty by reason of insanity. *American Journal of Psychiatry, 148*, 444–449.

Chiles, J., Von Cleve, E., & Trupin, E. (1990). Substance Abuse and Psychiatric Disorders in Prison Inmates. *Hospital and Community Psychiatry, 41*(10), 1132–1134.

Cirincione, C., Steadman, H., Robbins, P., & Monahan J. (1994). Mental illness as a factor in criminality: a study of prisoners and mental patients. *Crim Behav Ment Health, 4*, 33–47.

Clarke, G. N., Herincks, H. A., Kinney, R. F., Paulson, R., Cutler, D. L., Lewis K., & Oxman E. (2000). Psychiatric hospitalizations, arrests, emergency room visits, and homelessness of clients with serious and persistent mental illness: Findings from a randomized trial of two ACT programs vs, usual care. *Mental Health Services Research, 2*(3), 155–164.

Cutler, D. L. (1985). Clinical care update: the chronically mentally ill. *Community Mental Health Journal, 21*(1), 61–73.

Cutler D. L. (1986). Clinical Care Update: Community Residential options for the Chronically Mentally ill. *Com. Mental Health Journal, 22*(1), 61–73.

Cutler D. L., Tatum E., Shore J. H. (1987). A comparison of schizophrenic patients in different community support treatment approaches. *Community Mental Health Journal, 23*(2), 103–113.

Cutler D. L., Bigelow D., & McFarland B. (1992). The cost of fragmented mental health financing: Is it worth it? *Community Mental Health Journal, 28*(2).

Cutler D. L., McFarland B., & Winthrop K. (March 1998). Mental health in the Oregon health plan integration or fragmentation? *Administration and Policy in Mental Health, 25*(4), 361–386.

Field G., & Little R. (1998). Development of a program system for developmentally disabled inmates. *Offender Program Report*, 2(3), 33–45.

Field G. (1998). From the institution to the community. *Corrections Today*, October, 98–113.

Field G. (November 1998). Data relative to the Increases in mentally ill inmates entering prison from Multnomah County. Oregon Department of Corrections Memorandum.

Ganzini L., McFarland B., & Cutler D. (1990). Prevalence of mental disorders after catastrophic financial loss. *Journal of Nervous and Mental Disease, 178*, 680–685.

Goldfinger, S. M. (1990). Perspectives on the homeless mentally ill. *Community Mental Health Journal, 26*(5), 391–390.

Klassen D., & O'Connor W. (1988). A prospective study of predictors of violence in adult male mental patients. *Law Hum Behav. 12*, 143–158.

Kleinsasser D., West M., Michaud J., Hromas S., Gysin O., English K., Patrick D., Barrett T., Jenson K., Coen A., Befus J., Phelps E., Miller R., & Staples C. (1997). Prisoners with mental illness: A multi-agency task group report to the Colorado Legislature Joint Budget Committee.

Lamb, R. (1980). Structure: the neglected ingredient of community treatment. *Archives of General Psychiatry, 37*, 1224 1228.

Lamb, R. (1981). What did we really expect from deinstitutionalization? *Hospital & Community Psychiatry, 32*, 105–109.

Lamberti, J. S. (1999). Prevention of jail and hospital recidivism among persons with severe mental illness. Project Link, Department of Psychiatry, University of Rochester, Rochester NY. *Psychiatric Services, 50*(11), 1477–1480.

Larsell O. (1945). History of care of insane in the state of Oregon. *Oregon Historical Quarterly, 46*(4), 295–326.

Leff J., Dayson D., Gooch C., Thornicroft G., and Wills W. (1996). Quality of life of long stay patients discharged from two psychiatric institutions. *Psychiatric Services, 47*(1), 62–67.

Leopold W. S. (2000). A model prison diversion program—The criminal justice-community outreach department of the Montgomery county emergency service, Norristown Pa. *Psychiatric Services, 51*(11), 1440–1442.

London, Jack. (1982). "People of the Abyss" Star Rover House Oakland, California, pg. 270.

Masuda M., Cutler D., Hein L., & Holmes T. (1978 ). Life events and prisoners: A study of the relationship of life events to prison incarceration. *Archives of General Psychiatry, 35*(2), 197–203.

McFarland, B., & Bigelow, D. (1993). Chapter 11: Financial aspects of the Psychiatric Security Review Board. In J. Bloom, & M. Williams (Eds.), The management and treatment of insanity acquitees: A model for the 1990's. *Progress in Psychiatry Series*. Washington, D.C.: American Psychiatric Press, Inc.

Minkoff, K. (1987). Beyond deinstitutionalization: A new ideology for the postinstitutional era. *Hospital & Community Psychiatry, 38*, 945–950.

Munetz, M., Peterson, G., & Vandershie, P. (1996). SAFER houses for patients who need asylum. *Psychiatric Services, 47*, 117.

Munetz, M., Grande, T., & Chambers, M. (2001). The Incarceration of Individuals with severe mental disorders. *Community Mental Health Journal, 37*(4), 361–372.

Pattison, E. M., Llamas, R., & Hurd, G. (1979). Social network mediation of anxiety. *Psychiatric Annals, 9*, 474–482.

Petersilia, J. (1997). Justice for all? Offenders with mental retardation and the California corrections system. *The Prison Journal, 77*(4), 358–380.

Peterson, R., (1978). What are the needs of chronic mental patients. In *The chronic mental patient: Problems, solutions, and recommendations for policy*. Edited by John Talbott, Published by the American Psychiatric Association, 39–50.

Robins, L., & Regier, D. (Eds.) (1991). *Psychiatric disorders in America: The epidemiologic catchment area study*. New York: The Free Press (Macmillan, Inc.)

Rogers, J., Bloom, J., & Manson, S. (1986). Oregon's Psychiatry Security Review Board: A comprehensive system for managing insanity acquittees. In S. Shaw (Ed.), *Mental health and law: Research and policy*. Annals of the American Academy of Political and Social Sciences, *484*, pp. 86–98.

Skryha, V., Paulson R., & Eliot D. (1999). A look at supportive housing for mental health consumers in four Oregon Counties; Final Report (To Center for Mental Health Services) of Phase I of the Oregon Supportive Housing Evaluation Project (Grant No. 1 U1E SM52062–01).

Steadman, H., Morris, S., & Dennis, D. (1995). The diversion of mentally ill persons from jails to community based services: A profile of programs. *American Journal of Public Health, 85*(12), 1630–1635.

Stein, L. I., & Test, M. A. Alternative to mental hospital treatment. I. conceptual model, treatment program, and clinical evaluation. *Archives of General Psychiatry, 37*(4).

Strauss, J. S., Hafez, H., Lieberman, P., & Harding, C. M. (1985). The course of psychiatric disorder, iII: Longitudinal principles. *American Journal of Psychiatry, 142*(3), 289–296.

Teplin, L. (1990). The prevalence of severe mental disorder among male urban detainees: Comparison With the epidemiological catchment Area Program. *American Journal of Public Health, 80*(5), 663–669.

Tolsdorf, C. C. (1976). Social network support and coping. An exploratory study. *Journal of Family Process, 4*, 407–417.

Torrey, E. F, Bigelow, D., & Sladen-Dew. (1993). Quality and costs of services for seriously mentally ill individuals in British Columbia and the United States. *Hospital & Community Psychiatry, 44*, 943–950.

Treatment Advocacy Center (3/18/99). http://www.psychlaws.org.

Whitelaw, E. (Summer, 1997). What price prisons? *Oregonian Quarterly*, 20–24.

Part V

# PSYCHIATRIC ANTICIPATORY PLANNING

CHAPTER 10

# "CAN I PLAN *NOW* FOR THE MENTAL TREATMENT I WOULD WANT IF I WERE IN CRISIS?" OREGON'S PSYCHIATRIC ADVANCE DIRECTIVE

PATRICIA BACKLAR, BENTSON H. MCFARLAND, AND JO MAHLER

*The aspiration to make the goodness of a good human life safe from luck through the controlling power of reason.*
—MARTHA C. NUSSBAUM, *The Fragility of Goodness* (p. 3)

The poet tells us that Ulysses was warned in advance about seductive sirens whose enchanting melodies enticed seafarers to a rocky island shore near Scylla and Charybdis. These fabulous creatures promised voyagers wisdom but brought about their death. Ulysses directed his sailors to seal their ears with wax. He, however, longed to listen to the enthralling melodies, and commanded his crew to tie him to the mast. The verses of the sirens' song, which described a quickening of the spirit, were even more thrilling than the melody. Ulysses' heart ached with desire, but he could not break loose from his bonds and his immediate peril was safely avoided.

Advised about potential hazards, Ulysses prepared himself in advance and thus was able to conquer his impetuous passions. He was fortunate. Many of us yearn for order (Nussbaum, 1989) and struggle to control our lives, but rarely are we able to meet with such success (Backlar, 1995a).

Our interest in medical advance directives reflects not only our common human aspiration for order, but also a more specific hope to be able to protect ourselves and make our own health care decisions should we lose our capacity to make such determinations. Medical advance care planning became formalized with the implementation in 1991 of the Patient Self-Determination Act (Omnibus Budget Reconciliation Act of 1990). This Act, known as the PSDA, directs all health care institutions which receive federal funds to apprize patients, at the time of their admittance to an institution, of their rights under state law to prepare an Advance Directive (AD) for health care (Greco, Shulman, Lavizo-Mourey, & Hansen-Flaschen, 1991). An AD is a legal document that allows people to specify the medical treatment they wish to obtain if they should, at a future time, lack the capacity to make their own health care decisions.

Over the years, debates have been waged about the feasibility of contracts, made in advance, for and against psychiatric treatment (Appelbaum, 1991). Szasz (1982) challenged the imposition of involuntary psychiatric treatment; he promoted the idea of a "psychiatric will," e.g., competent persons who wish to refuse involuntary psychiatric treatment should attempt to make their choices legally binding with the use of a living will. On the other hand, Rosenson and Kasten (1991) proposed a contract that would confirm consent to psychiatric treatment; when competent persons make a prior agreement that binds them to future treatment, these agreements may be referred to as Ulysses contracts (Dresser, 1984). A centrist recommendation, made by Rogers and Centifanti (1991), allows competent individuals to state their own particular preferences for acceptance or rejection of psychiatric treatment. Supplementary or parallel to these substantive proposals are procedures that permit competent persons to appoint agents—surrogate decision-makers—to represent them should they lose their capacity to make their own medical and psychiatric determinations (Appelbaum, 1991; NY Pub Health Law, 1991; OBRA, 1990; President's Commission, 1983; Rogers & Centifanti, 1991).

In many ways, psychiatric advance directives (PADs) have been modeled upon ADs for end-of-life care. Both allow autonomous persons to plan ahead for a time when decisionmaking capacity may be impaired, and to put in place protections of their own choosing (Backlar,

1995a; Backlar, 1997; Backlar and McFarland, 1996, 1998). Yet, the two kinds of directives differ in substance. The AD for end-of-life mainly addresses circumstances immediately preceding a singular event—the principal's death. In contrast, the PAD concerns events that may occur repeatedly, resulting from an ongoing condition and fluctuating incapacity to make mental healthcare decisions. Moreover, PADs are intended for persons who already have experienced the sort of crisis that they anticipate will recur; who will, thus, be able to use such experience to plan for similar situations in the future, or perhaps even prevent them (Backlar, 1997). In short, while ADs prepare for dying, PADs are concerned with restoration, recovery—with living (Lefley, 1996).

## THE OREGON DOCUMENT

Provisions for advance planning for psychiatric treatment were formulated by the 67th Oregon Legislative Assembly, which enacted Senate Bill 859 in 1993. The Oregon "Declaration for Mental Health Treat-ment" (Oregon Revised Statutes, 1993) was among the first documents of its kind in the United States. Fourteen states now have laws that specify a legal PAD (Stavis, 2000), and according to Fleischner (1998), forty-six states permit some psychiatric treatment to be addressed in generic ADs.

The Oregon PAD is both substantive and procedural. Substantively, the legal document authorizes competent persons to prospectively make decisions about what mental health treatment they would or would not want. Principals may delineate and detail precisely their treatment preferences in regard to psychoactive medications, electro-convulsive treatment, and to being confined in a mental health facility for up to 17 days. However, there is a caveat: whether or not persons have signed a legal "Declaration," if two physicians have put them on an emergency psychiatric hold, or if they have been committed by a court, a physician still may prescribe medications, under strict legal guidelines, that are against a principal's documented (or undocumented) wishes (Backlar, Asmann, and Joondeph, 1994).

Procedurally, surrogate decisionmakers may be appointed. All such agents must confirm that they agree to act as the principal's representative by signing the legal document. Agents are expected to adhere to the principal's written or spoken competently-made determinations or, if that information is not known to them, to act in the principal's best interests. The appointed agent may not make psychiatric treatment decisions unless the principal is incapable. While the principal remains incapable, the Oregon PAD allows a limited waiver of confidentiality that permits healthcare providers to discuss the principal's psychiatric treatment information with the appointed agent. Even though agents are not personally held responsible for the cost of the principal's treatment, they may have to provide evidence to a provider that a payment source exists before treatment or admission occurs (Barret, 1995). In order to protect the principal against coercive influences, health care providers, owners of a treatment facility, or their relatives or employees, in which the principal is a patient or resident, may not serve as surrogate decisionmakers (unless they are related by blood, marriage, or adoption to the principal). The "Declaration" is validated by two witnesses who must sign the document, confirm that they personally know the principal, and that they believe the principal to be competent at the time that the principal has signed the form. Precisely the same limitations hold for the witnesses as for the surrogate decisionmakers, with the added constraint that witnesses may not be related to the principal.

A completed "Declaration" remains valid for three years. Competent principals may revoke the document at any time. However, if they are judged to be in an incapacitated state by two physicians or by a court in a guardianship proceeding, they may not make any alterations to the document during the period of time that they are deemed to lack capacity for making such mental health treatment decisions (Barret, 1995; Oregon Revised Statutes, 1993).

The Oregon State Mental Health Division adopted an Acute Care Administrative Rule (OAR, 1994) asserting that all persons with severe and persistent mental disorders must be notified about the existence of the PAD at the time of hospital admission or discharge. At the time of this writing, the State Mental Health Division is in the process of adopting an Adult Care Administrative Rule which will require providers in the community mental health centers to inform all consumers about their rights to prepare an Oregon PAD, and to offer assistance—if requested—with PAD preparation.

## PADs MAY SERVE INSTRUMENTAL AND MORAL FUNCTIONS

The plausible instrumental functions of a PAD may reduce hospitalizations, civil commitments, and use of court appointed guardians. The possible moral functions are more amorphous: Consumers may feel empowered and less anxious about their treatment should future crises occur; and relationships between consumers and families, consumers and providers, and families and providers may be improved.

## Potential Benefits

In the best of all possible worlds, when consumers have capacity to understand and discuss treatment information over a period of time with their providers (Appelbaum & Grisso, 1988; Katz, 1984; Lacro, Sewell, Warren, Woody, Harris, & Jeste, 1994) and their family members (if such persons are willing), the use of an instrument like an advance directive may strengthen the informed consent process. In many ways the functions of informed consent and anticipatory planning serve similar purposes. Both processes: show respect for persons by promoting their autonomy and self-determination; empower persons to communicate and protect their own interests; and regulate and balance the relationship between patients and providers (Capron, 1991). If such documents are incorporated into informed consent procedures they may help providers: to engage the client to participate in her/his own treatment plan; to discuss medication and treatment options with the client; to judge, through ongoing dialogue, how well the client can make appropriate decisions (Miller, 1998); to listen carefully to the client's wishes and values, and evaluate whether the client's goals are realistic; to cite illustrations from the client and/or the provider's experience that illuminate what might or might not work; to get to know and regularly talk to the appointed surrogate decision-maker (taking into account that this individual eventually may be making decisions for the client); and, to routinely review the advance directive with the client (Pinkney, 1991).

There has been speculation that the whole community may benefit, both procedurally and financially, if trusted surrogate decision-makers are appointed. For instance, treatment decisions might be made in a timely fashion, affording consumers preventive care and averting coercive measures (Lucksted & Coursey, 1995). And, some hospitalizations, involuntary commitments, appointment of legal guardians, and needs for court proceedings may be lessened or avoided (Herr & Hopkins, 1994; The English Law Commission, 1991).

Some commentators suggest that by offering a consumer choice and control, the PAD may be itself a therapeutic mechanism, e.g., Wexler (1994) considers such "therapeutic jurisprudence" as a means for "bringing mental health insights into the development of the law" (p. 259). In an article on patients' rights to refuse or choose mental health treatment, Winick (1994), noting that law may work as a rehabilitative instrument, emphasizes the psychological value of choice. He supports the view that patients' choice is an important determinant of treatment success. "Patients," he writes, "like people generally, often do not respond well when told what to do" (Winick, 1994, p. 100).

## Potential Obstacles

Other writers consider that PADs may not be suitable for all consumers. Even when their disorders are under control, some consumers' reasoning abilities and insight may be compromised. Torrey (1995) writes "40–50 percent of seriously mentally ill individuals have significantly and permanently impaired insight into their illness." Making choices, which may be acted upon at a future time, may be too abstract for some consumers.

Appointing a surrogate decisionmaker may be a difficult task for some consumers. According to a 1992 national survey (Skinner, Steinwachs, & Kasper, 1992) of 1,401 member families of the National Alliance for the Mentally Ill (NAMI), almost 59% of persons with severe and persistent mental disorders did not live with their families (Lefley, 1996). Many persons with schizophrenia have a limited support system and few or no relationships (Cutler, Tatum, & Shore, 1987). For some persons their provider may be their only trusted friend and they may wish to select her/him as their agent. But this appointment is forbidden by the statute.

All anticipatory planning—whether for medical or psychiatric treatment—may have a fundamental defect. What we wish for while we are in one mental state may no longer be what we want or need when we are in an altered mental state (Dresser, 1994; Backlar, 1995a; Howe, 2000). The interests of a competent person may change profoundly when the same person no longer has the capacity to make health care decisions (Robertson, 1991). "One powerful objection to advance directives is that they require individuals to make decisions in the present about conditions which may or may not arise in the future and which the individual may never have experienced" (Hornett, 1995). The Oregon PAD does not permit revocation while the principal is incompetent. As noted above, whether or not a person has made out a PAD, if s/he is on an emergency psychiatric hold, or has been committed by a court, a physician—under strict legal guidelines—may still prescribe medicine that does not comply with the consumer's instructions. This may undermine the principal's confidence in the PAD.

The method, used in Oregon, for verifying the principal's capacity to prepare a PAD may be imperfect. The two witnesses, whose signatures are meant to affirm the principal's capacity, may themselves not have the capacity to assess the principal's competence. Such informal witnessing is standard procedure in the execution of wills and medical ADs. But when people are making mental health decisions of significant consequence, it may be prudent to institute a more formal process that can confirm principal's capacity to execute a PAD.

Systems efficiencies and cost savings may not occur. If consumers use the PAD to assert their negative rights to refuse treatment, hearings to establish competency may continue to be employed. On the other hand, requests based on positive rights to treatment also may be ignored. As Winick (1996) notes, the right to refuse treatment is fundamental, whereas the right to demand treatment is not similarly protected. Consumers who explicitly request the newer antipsychotic medications, like clozapine, risperidone, or olanzapine, while refusing medications like chlorpromazine or haloperidol, which have unpleasant and deleterious side effects, may find that such requests are denied.

Even though the Oregon document allows a principal to request voluntary hospitalization (up to 17 days), in Oregon beds are rarely available. Indeed, throughout the U.S., the combination of state hospital downsizing, managed mental health care, and the blending of public and private sectors has accelerated restrictions on hospitalizations and multiplied premature discharge of even severely psychotic persons (Lefley, 1996). Requests by consumers, who are in crisis, for voluntary inpatient treatment may be ignored.

Overburdened providers of mental health services may balk at the additional work imposed as a consequence of helping consumers prepare PADs. At a time when resources are scarce, the future of federal entitlement programs for people with severe and persistent mental disorders is uncertain, and ways of delivering mental health care services are in flux (in part due to the advent of managed care systems that are geared more toward providing acute services for acute short-term ailments rather than chronic long-term conditions), funding for time spent on such activities may not be available (Backlar, 1995b). Furthermore, some providers may perceive PADs as another unwanted intrusion of the law into mental health care practice (Hughes & Singer, 1992).

## OPINIONS ABOUT PADs IN OREGON

A study, "Impact of advance directives for mental health treatment," the first in the US to obtain patient, provider, and surrogate opinions about

PADs (Backlar, McFarland, Swanson, & Mahler, in press), was undertaken subsequent to an informal survey of PADs (Backlar & McFarland, 1996; 1998), four years after the Oregon document was legalized.

## METHODS

The study drew on a convenience sample of eligible adult consumers (*N*=40) with serious mental illness (SMI) who were receiving treatment in public mental health programs. Approximately 251 adult consumers with SMI, who came in for appointments, were given information about the Oregon PAD by their providers and told about the study. The first 40 consumers who agreed to be in the study comprised the study sample. Collateral informants included consumers' providers (*N*=21) and consumers' appointed surrogate decisionmakers (*N*=17). Some providers gave collateral information for more than one consumer.

Interviews with consumers took, on average, about 50 minutes. Baseline and follow-up interviews (8–10 months later) were conducted face-to-face. Surrogate and provider interviews were conducted by telephone. The interviews consisted mainly of open-ended questions. Topics addressed in the questionnaires included: capacity to prepare a PAD, surrogate appointment, PAD satisfaction, treatment choices elected, PAD impact on consumer/provider/surrogate relationships and on consumers' attitudes to treatment, and respondents opinions about placing PAD information in a state computerized database.

Thirty consumers agreed to prepare a PAD, while 10 consumers refused to prepare the document. As reported by their providers, the majority of consumers—21 consumers (70%) in the PAD group, and 7 consumers (70%) in the no-PAD group—had a diagnosis of schizophrenia. There were more women (63.3%) in the PAD group, and more men (70%) in the no-PAD group. In the PAD group, more persons had been hospitalized in the past year: eight consumers (27.6%) in the PAD group compared with only one subject (10%) in the no-PAD group (see Table 1 for patient demographics and clinical characteristics). Policy

TABLE 1. Consumer Demographics and Clinical Characteristics Based on Consumer (Except where noted) Responses

| Responses | Completed AD | No AD | Total |
|---|---|---|---|
| *Sample Size* | 30 (75.0%) | 10 (25.0%) | 40 (100.0%) |
| *Gender* | | | |
| Female | 19 (63.3%) | 3 (30.0%) | 22 (55.0%) |
| Male | 11 (36.7%) | 7 (70.0%) | 18 (45.0%) |

TABLE 1. Continued

| Responses | Completed AD | No AD | Total |
|---|---|---|---|
| *Race* | | | |
| White | 29 (96.7%) | 9 (90.0%) | 38 (95.0%) |
| Asian-Amer. | 0 | 1 (10.0%) | 1 (2.5%) |
| Other | 1 (3.3%) | 0 | 1 (2.5%) |
| *Mean age in years (sd)* | 39.73 (8.44) | 36.0 (12.02) | 38.80 (9.43) |
| [1]Dx | | | |
| Schizophrenia | 21 (70.0%) | 7 (70.0%) | 28 (70.0%) |
| Other | 9 (30.0%) | 3 (30.0%) | 12 (30.0%) |
| *Years of education* | | | |
| 8–11 | 6 (20.0%) | 1 (10.0%) | 7 (17.5%) |
| 12 | *15 (50.0%) | 3 (30.0%) | 18 (45.0%) |
| 13–15 | 6 (20.0%) | 4 (40.0%) | 10 (25.0%) |
| 16 or more | 3 (10.0%) | 2 (20.0%) | 5 (12.5%) |
| | *Ged 2 subjects | | |
| *Marital status* | | | |
| Single | 28 (93.3%) | 10 (100.0%) | 38 (95.0%) |
| Married | 2 (6.7%) | 0 | 2 (5.0%) |
| *Appointed surrogate* | | | |
| Yes | 22 (73.3%) | N/A | N/A |
| No | 8 (26.7%) | N/A | N/A |
| [1]*Psychiatric hosp.* | | | |
| [2]Ever | | | |
| Yes | 29 (96.7%) | 10 (100.0%) | 39 (97.5%) |
| No | 1 (3.3%) | 0 | 1 (2.5%) |
| In past year | | | |
| Yes | 8 (27.6%) | 1 (10.0%) | 9 (23.1%) |
| No | 20 (69.0%) | 9 (90.0%) | 29 (74.4%) |
| Don't know | 1 (3.4%) | 0 | 1 (2.6%) |
| *Court commitment* | | | |
| Ever | | | |
| Yes | 14 (46.7%) | 5 (50.0%) | 19 (47.5%) |
| No | 16 (53.3%) | 5 (50.0%) | 21 (52.5%) |
| In past year | | | |
| Yes | 2 (14.3%) | 0 | 2 (10.5%) |
| No | 11 (78.6%) | 5 (100.0%) | 16 (84.2%) |
| Unsure/Don't know | 1 (7.1%) | 0 | 1 (5.3%) |
| *Arrested* | | | |
| Ever | | | |
| Yes | 14 (46.7%) | 5 (50.0%) | 19 (47.5%) |
| No | 16 (53.3%) | 5 (50.0%) | 21 (52.5%) |
| Past Year | | | |
| Yes | 1 (7.1%) | 0 | 1 (5.3%) |
| No | 13 (92.9%) | 5 (50.0%) | 18 (94.7%) |

[1]Based on provider responses.
[2]Note: All consumer, provider and surrogate responses were congruent.

makers will be interested learn whether further, and larger, studies also demonstrate that consumers who have recently experienced a hospitalization are more likely to prepare a PAD.

## Results

The majority of responding consumers (23 of 28, or 82.1%) in the PAD group had capacity to understand many of the PAD concepts, and their collateral informants (94.1% of responding surrogates, and 100% of responding providers) verified this impression. Of the 30 consumers who completed a PAD, 22 (73.3%) appointed a surrogate decision-maker, and in most cases, consumers (19 of 22, or 86.4%) selected a close family member to be their representative. Similar to findings from an earlier informal survey (Backlar & McFarland, 1996, 1998), almost one-third of consumers did not appoint a surrogate.

No one used the PAD to refuse all treatment, although specific treatments were refused. Six of 30 consumers chose to defer to their physician's treatment decisions, while 6 of 30 respondents asked for their current treatment to be continued. Seventeen of 30 consumers specifically refused shock treatment, and 8 respondents refused haloperidol. Because the consumer participants were self selected, the finding that they did not use the PAD to refuse all treatment may be considered suspect. Nonetheless, in the general population persons who decide to prepare a PAD will also be self-selected—individuals will be invited, not forced, to prepare a PAD.

The majority of all respondents said they would approve if the state placed PADs in a computerized database. Among providers who agreed that there was a need for a central registry, 14 of 21 were concerned about the potential for breaching confidentiality. Only five of the 30 consumer respondents mentioned their concern about confidentiality. If states choose to establish PAD registries, a precedent already exists in the national registry that tracks blood tests required of mentally ill persons who are prescribed the medication clozapine (Honigfeld, 1996).

PADS were acceptable to many respondents. Twenty-six of 30 responding consumers (86.7%), and all 17 responding surrogates (100%) approved of the PAD. Most responding providers were satisfied with the manner in which their clients (28 of 30, or 93.3%) prepared the PAD. Yet, in general, most responding providers (11 of 15, or 80%) evinced serious concern about PAD implementation. They spoke of having "little confidence that institutions involved [would] have a system for accessing the [information]," and suggested that the mental health system find a way to make the PAD be "more a standard operating procedure, so that people would automatically have them."

PAD preparation made consumers (85.2%) feel empowered. Yet, their responses were revelatory. Their remarks exposed how vulnerable they feel to the exigencies of their mental illness, and also the extent to which they perceive themselves to be at the mercy of the mental healthcare system. Indeed, consumers' initial enthusiasm for the PAD may simply be indicative of how powerless they feel.

The 10 consumers who did not prepare a PAD gave various reasons for declining. One individual decided to "trust that I wouldn't get ill again." A few consumers felt they that their doctors, or the system, could be counted upon to look after them. One consumer didn't want "anyone to make decisions for me." However, 6 consumers indicated it was too difficult for them to put the PAD together: "it wasn't talked about enough"; "I hadn't heard about it"; My brother wouldn't sign it—he just let it ride." These responses reconfirm findings from an earlier informal survey (Backlar & McFarland, 1996, 1998) that without sufficient provider education, and direct assistance and support for consumers, it is unlikely that many consumers will prepare a PAD.

## Follow-up Interviews

Participants in follow-up interviews included 26 of the 30 consumers who prepared a PAD, 8 of the 17 surrogates originally interviewed, and 17 providers who answered questions pertaining to 24 of the 26 consumers participating in the second round of interviews.

Satisfaction with the PAD was reconsidered. The majority of responding consumers (88.5%) were still satisfied, but their responses to open-ended questions lacked the enthusiasm they had expressed in their baseline interview. Indeed, a number of consumers (46.2%) were concerned that providers had been inadequately trained about PADs and that the PAD might ignored in a crisis situation. Providers (who reported about 17 of their clients) said that the PAD would be adequate for these client's needs and that it was a "positive influence," Yet, in general, a majority of responding providers (64.3%) reiterated their baseline interview concerns about the lack of PAD standards and procedures, and the need to have a PAD registry in order to "figure out a way to remember [that their clients' PADs] exist."

The majority of providers (71.4%) said that PAD preparation had not shifted their clients' views—one way or another—about taking medication. The majority of responding surrogates (62.5%) also believed that the PAD had little effect in regard to consumers' medication attitudes. Although some providers' (28.6%) responses indicated that the PAD may be a constructive tool for a few of their clients who participated in the research, this small study appears to demonstrate that the PAD is a modest tool in regard to influencing patients' attitudes—either positively or negatively—toward medication or in regard to their relationships with their providers.

During the 8–10 month period following the baseline interviews, providers reported no hospitalizations for five subjects in the no-PAD group, and no information was given for the other five no-PAD consumer participants. In the PAD group, providers reported that three of 24 consumers had been hospitalized (one person was hospitalized twice). Only one of these four hospitalizations was involuntary. In the case of the consumer participant who was hospitalized twice, the outpatient provider and the consumer made the hospital aware of the PAD, and discussed the document with the inpatient clinicians.

## A VIGNETTE

The following vignette about a hospital experience—recounted by a research participant, his surrogate, and his outpatient provider—illustrates how a legal PAD may be ignored by inpatient and outpatient clinicians, when there is virtually no policy, and administrative rules are not in place, to serve as guide to the providers.

The research participant reported discussing the PAD with the attending physician, "I wanted them to know I had prior wishes before I came to the hospital." The inpatient providers, however, ignored the PAD. "I don't know [what] healthcare workers ... think about [the PAD], but if they don't back it up, it's not useful," the participant said. He indicated that he felt as though he was not respected, "I hope that those [responsible for] mental health treatment become a bit kinder in their methods of helping clients ... ."

The participant's surrogate explained the circumstances surrounding the her relative's attempts to get psychiatric treatment: "He had such a challenge getting hospitalized ... the social worker was not available ... he had paranoid symptoms ... ran out of the prescription and the pharmacy would not refill ... ." The surrogate was reluctant to mention the PAD to the hospital staff: "I was afraid [that the inpatient physician] would think 'not only are you trying to get admitted but you also want to treat yourself.'" There was also an issue of not being able to communicate with the outpatient provider, "When [my relative] tried to call [from the hospital], I was present. Someone said that the provider was not available. The substitute for the provider did not know the case and didn't respond." Despite being the patient's legally authorized representative (surrogates may assume decisionmaking responsibilities when patients are incapable, at such times the legal PAD permits a limited waiver of confidentiality) the surrogate did not participate in the treatment decisionmaking process—"no one asked me."

Ironically, the consumer participant had stated in the baseline interview that if a hospitalization occurred the surrogate could be trusted to "look after my best interests—to monitor my concerns as I am being treated." Some might argue that the surrogate should have made a greater effort to speak out and ensure that the consumer's PAD was at least reviewed. It has long been recognized, however, that patients and their relatives feel vulnerable and dependent, even in commonplace encounters with their physicians (Parsons, 1951).

The outpatient provider confirmed that the PAD "never came into play during [the recent] hospitalization," and acknowledged there had been no involvement or communication with the client's surrogate decision-maker or with the inpatient clinicians. "[I] found out pretty much after the fact—no contact with [the hospital staff]."

## NEXT STEPS

It is important to recognize that the poor communication between the stakeholders (as described in this vignette) is not entirely due to the research participant's PAD being ignored. Many other factors are in play. Indeed, the results of this study appear to substantiate that a piece of paper by itself may not alter (one way or another) consumer attitudes, improve outcomes, remedy a lack of resources, or compel clinicians to be kinder to their patients.

Originally, the PAD was conceived as a self-advocacy tool for consumers to use in psychiatric emergencies (Backlar, 1995; Sutherby, Szmukler, Halpern, Alexander, Thornicroft, Johnson, & Wright, 1999). PADs, however, may turn out primarily to be a mechanism in which a consumer's authentic consent (prepared in advance) to a treatment plan, which is intended to take effect in the event of the consumer's decompensation, or crisis (and made in collaboration with her/his provider), is chronicled and, when required, easily retrieved. Such a mechanism, may facilitate providers' ability to more effectively orchestrate continuity of care for persons with serious mental illness. If PADs encourage stakeholder communication, that could be an important secondary effect.

In general, a legal change—like a PAD—is a necessary but insufficient step toward creating social or political change (Durham, 1999). Indeed, the descriptive findings in this study may be interpreted to indicate that PAD legislation alone does not translate into adequate policy. Although study respondents (consumers, providers, and surrogates) found the PAD acceptable, I suspect that without supplemental supportive enhancements—like intensive stakeholder education, assistance for consumers in PAD preparation, a central registry where PADs may be documented and easily retrieved when necessary (clearly, further studies are needed to examine the value of such interventions)—the Oregon PAD may easily be ignored by clinicians in outpatient and inpatient treatment facilities.

Realistically, and despite the potential for therapeutic jurisprudence (Backlar and McFarland, 1998; Wexler, 1994; Winnick, 1994), the PAD is likely to be a modest tool with regard to patient outcomes. Nevertheless, in a fragmented treatment system complicated both by disparate treatment locations and by providers who—usually through no fault of their own—may not be available when needed (as was illustrated in the vignette), the PAD's most tangible significance may be as an instrument that assists providers to enhance consumers' continuity of treatment and accordingly improve the opportunity for consumer rehabilitation and recovery (Backlar, 2001; Backlar, McFarland, Swanson, & Mahler, 2001; Halpern & Szmukler, 1997).

ACKNOWLEDGEMENTS. This paper was partially supported by National Institute of Mental Health grant R03MH55969 to Patricia Backlar.

A version of this paper was previously published in *Administration and Policy in Mental Health*, 2001.

## REFERENCES

Appelbaum, P. S. (1991). Advance directives for psychiatric treatment. *Hospital and Community Psychiatry, 42*, 983–984.

Appelbaum, P. S. (1994). *Almost a Revolution: Mental Health Law and the Limits of Change.* New York: Oxford University Press.

Appelbaum, P. S., & Grisso, T. (1988). Patients' capacities to consent to treatment. *New England Journal of Medicine, 319*, 1635–1638.

Backlar, P. (1994). *The Family Face of Schizophrenia.* New York: Tarcher/Putnam.

Backlar, P. (1995a). The longing for order: Oregon's medical advance directive for mental health treatment. *Community Mental Health Journal, 31*, 103–108.

Backlar, P. (1995b). Health care reform: Will the subject fall out of the topic? *Community Mental Health Journal, 31*, 297–301.

Backlar, P. (1997). Anticipatory planning for psychiatric treatment is not quite the same as planning for end-of-life care. *Community Mental Health Journal, 33,* 261–268.

Backlar, P. (2001). Privacy and confidentiality. In *Textbook of Community Psychiatry,* (Eds.) Graham Thornicroft and George Szmukler, Oxford University Press.

Backlar, P., Asmann, B. D., & Joondeph R. C. (1994). *Can I plan now for the mental health treatment I would want if I were in crisis? A Guide to Oregon's Declaration for Mental Health Treatment.* Salem, Oregon: Office of Mental Health Services, Mental Health and Development Disability Services Division.

Backlar, P., & McFarland, B. H. (1996). A survey on use of advance directives for mental health treatment in Oregon. *Psychiatric Services,* 47, 1387–1389.

Backlar, P., & McFarland, B. H. (1998). Oregon's Advance directive for mental health treatment: Implications for policy. *Administration and Policy in Mental Health, 25,* 609–618.

Backlar, P., McFarland, B. H., Swanson, J. W., & Mahler, J. (2001). Consumer and provider views on psychiatric advance directives. *Administration and Policy in Mental Health,* 28, 427–441.

Barret C. L. (1995). Practical comments on the Advance Declaration for Mental Health treatment. *Oregon Estate Planning and Administration Section Newsletter, XII*: 1–2.

Brown, S. J. (1995, July 10–11). "An idea whose time has come": Advance directives move into mental health care. *Clinical Psychiatry News,* pp. 10–11.

Callahan, D. (1995). Terminating life-sustaining treatment of the demented. *Hastings Center Report, 25*(6), 25–31.

Capron, A. M. (1991). Protection of research subjects: Do special rules apply to epidemiology? *Law, Medicine, and Health Care, 19*(3–4), 184–190.

Cutler, D. L., Tatum, E., & Shore, J. H. (1987). A comparison of schizophrenic patients in different community support approaches. *Community Mental Health Journal, 23,* 103–113.

Dresser, R. (1984). Bound to treatment: The Ulysses contract. *Hastings Center Report, 14,* 13–16.

Durham, M. L. (1999). Personal communication.

Fleischner, R. D. (1998). An analysis of advance directive statutes and their applications to mental health care and treatment. Prepared for the Advocacy Training/Technical Assistance Center (ATTAC) of the National Association of Protection & Advocacy Systems, Inc. (NAPAS).

Greco P. J., Shulman, K. A., Lavizzo-Mourey, R., & Hansen-Flaschen, J. (1991). The patient self-determination act and the future of advance directives. *Annals of Internal Medicine, 115,* 639–643.

Halpern A., & Szmukler, G. (1997). Psychiatric advance directives: reconciling autonomy and non-consensual treatment. *Psychiatric Bulletin, 21,* 323–327.

Herr S. S., & Hopkins, B. L. (1994). Health care decision making for persons with disabilities: an alternative to guardianship. *Journal of the American Medical Association, 271,* 1017–1022.

Honigfeld, G. (1996). Effects of the clozapine national registry system on incidence of deaths related to agranulocytosis. *Psychiatric Services, 47,* 52–55.

Hornett, S. (1995). Advance Directives. In J. Keown, (Ed.), *Euthanasia Examined: Ethical, Clinical and Legal Perspectives.* Cambridge: Cambridge University Press.

Howe, E. G. (2000). Commentary on "Psychiatric advance directives: An alternative to coercive treatment." *Psychiatry, 63,* 171–177.

Hughes, D. L., & Singer, P. A. (1992). Family physicians' attitudes toward advance directives. *Canadian Medical Association Journal, 146,* 1937–1944.

Katz, J. (1984). *The Silent World of Doctor and Patient.* New York: Free Press.

Lacro, J. P., Sewell, D. D., Warren, K., Woody, S., Harris, M. J., & Jeste, D. V. (1994). Improving documentation of consent for neuroleptic therapy. *Hospital and Community Psychiatry, 45*, 176–178.

Lefley, H. P. (1996). *Family Caregiving in Mental Illness*. Thousand Oaks, California: Sage Publications.

Lucksted, A., & Coursey, R. D. (1995). Consumer perceptions of pressure and force in psychiatric treatments. *Psychiatric Services, 46*, 146–150.

Miller, R. D. (1998). Advance directives for psychiatric treatment: A view from the trenches. *Psychology, Public Policy, and Law, 4*, 728–745.

Nussbaum, M. C. (1986). *The Fragility of Goodness*. Cambridge: Cambridge University Press. pp. 3, 6.

Nussbaum, M. C. (1989). Recoiling from reason. *The New York Review of Books, XXXVI* (19), 36–41.

New York Pub Health Law, (1991). Art 29-C.

Omnibus Budget Reconciliation Act of 1990 (OBRA 1990); formerly this Act was called the "Patient Self-Determination Act of 1990." Sec. 4206: Medicare Provider Agreements Assuring the Implementation of a Patient's Right to Participate In and Direct Health Care Decisions Affecting the Patient. Public Law no. 101–508.

Oregon Administrative Rule, 1994, 309–34–870 [3] [1].

Oregon Revised Statutes, 1993, ORS 127.700–127.737.

Parsons T. (1951). The sick role and the role of the physician. *Milbank Memorial Fund Quarterly, 53*, 257–78.

Pinkney, D. S. (1991, November 25). Facilities must ask about advance directives: law may encourage talk about end-of-life care. *American Medical News*, pp. 28–29.

President's Commission for the Study of Ethical Problems in Medicine and Behavioral Research (1983). Deciding to Forego Life-Sustaining Treatment: Ethical, Medical, and Legal Issues in Treatment Decisions. Washington, DC, US Government Printing Office.

Robertson, J. A. (1991). Second thoughts on living wills. *Hastings Center Report, 21*, 6–9.

Rogers, A. J., & Centifanti, J. B. (1991). Beyond "self-paternalism": Response to Rosenson and Kasten. *Schizophrenia Bulletin, 17*, 9–14.

Rosenson, M. K., & Kasten, A. G. (1991). Another view of autonomy: Arranging consent in advance. *Schizophrenia Bulletin, 17*, 1–17.

Skinner, E. A., Steinwachs, D. M., & Kasper, J. D. (1992). Family perspectives on the service needs of people with severe and persistent mental illness. *Innovations and Research, 1*, 23–34.

Sutherby, K., Szmukler, G. I., Halpern, A., Alexander, M., Thornicroft, G., Johnson, C., & Wright, S. (1999). A study of crisis cards' in community psychiatric service. *ACTA Psychiatrica Scandinavaica, 100*, 56–61.

Szasz, T. (1982). The psychiatric will: A new mechanism for protecting against "psychosis" and "psychiatry." *American Psychologist, 37*, 762–770.

The English Law Commission (1991): Consultation Paper No. 119: Mentally incapacitated adults and decision-making: An overview. London, England: Her Majesty's Stationary Office.

Torrey, E. F. (1995, August 9). Personal communication.

Wexler, D. B. (1994). An orientation to therapeutic jurisprudence. *New England Journal on Criminal and Civil Confinement, 20*, 259–264.

Winick, B. J. (1994). The right to refuse mental health treatment: A therapeutic jurisprudence analysis. *International Journal of Law and Psychiatry, 17*, 99–117.

Winick, B. J. (1996). Advance directive instruments for those with mental illness. *University of Miami Law Review, 51*, 57–95.

CHAPTER 11

# PSYCHIATRIC HEALTH CARE PROXIES IN MASSACHUSETTS: MUCH TO DO ABOUT NOTHING, SO FAR

JEFFREY GELLER

*Life is what happens to you*
*while you're busy making*
*other plans.*
—JOHN LENNON (1980, 81)

Twenty years ago, Thomas Szasz indicated, "competent American adults should have a recognized right to reject involuntary psychiatric interventions that they may be deemed to require *in the future*, when they are *not competent* to make decisions concerning their own welfare" (Szasz, 1982, italics in original). About ten years later, Rosenson and Kasten (1991) opined, "the most authentic expression of autonomy may be the decision by a patient whose psychiatric symptoms are in remission to plan for treatment in the event of a crisis." At the same time, the Supreme Court of New York observed, "the fundamental right of individuals to have final say in respect to decisions regarding their medical treatment extends equally to mentally ill persons who may not be treated as persons of lesser status or dignity because of their illness."

The court ruled, "Absent an overriding State interest, a hospital or medical facility must give continued respect to a patient's competent rejection of certain medical procedures even after the patient loses competence" (Matter of Rosa M, 1991).

During this same time period, the general medical field was grappling with the issues of advance directives and health care proxies. The United States Supreme Court reached a highly publicized "right to die" decision in the Cruzan case (1990) and the United States Congress passed the Patient Self-Determination Act (PSDA) effective December 1, 1991.

In the wake of the PSDA, the American Psychiatric Association Board of Trustees approved a resource document, "The Patient Self-Determination Act: What Every Psychiatrist Should Know" in December, 1992 (APA Division of Government Relations). With all this attention to patients' capacity to influence their future treatment, Appelbaum (1991) commented in his column in *Hospital and Community Psychiatry* that "Advance directives for psychiatric treatment—long-debated, but little used—are on the verge of having a major impact on psychiatric care." Has that happened?

## MASSACHUSETTS HEALTH CARE PROXY

Massachusetts passed a health care proxy statute in 1990. The person who executes the document (referred to as the principal) indicates in writing who can act on his/her behalf (the agent) when he/she lacks the capacity to make health care decisions and what, if any, limits he/she wants to place on the agent's authority. The "capacity to make health care decisions" is clearly defined and is assumed until the attending physician determines, and indicates in writing, that the principal lacks the capacity to make or communicate health care decisions. If the incapacity is due to mental illness, the opinion of a psychiatrist must be sought (Cross, Fleischner, & Elder, 1994).

While the agent acts on the principal's behalf once a physician indicates the principal cannot make or communicate informed decisions, the principal may revoke the proxy at any time. This is particularly important in the psychiatric setting. Even after a physician has determined a patient to be incompetent, the patient can revoke any advance directive because the law provides for the presumption of competency until a court decides otherwise. Further, a patient who objects to his/her agent's decision, even after a physician determines him/her to be incompetent, must have that objection honored, because the law indicates that a principal's wishes shall prevail over the agent's decision unless a court decides the principal lacks the capacity to make health care decisions (Cross et al., 1994).

How does all this play out? There are two areas in psychiatric treatment where health care proxies appear to be most salient: admission decisions and treatment (medication) decisions.

**Admissions**. A patient shows up in a general hospital emergency room for psychiatric admission. If the patient meets the criteria for inpatient level of care (medical necessity) and agrees to be admitted, he is admitted. If the patient meets commitment criteria, he can be admitted independent of his expressed desire. If the patient has previously been ejudicated incompetent and has a guardian, the guardian cannot admit his ward over the ward's refusal, but must get a court order to admit. However, the clinically incompetent person can sign in voluntarily.

If the patient shows up with a health care proxy and his agent, the agent could sign the patient in if the physician determines the patient is not competent and if the patient does not object. If the patient indicates he does not want to be admitted, the agent's decision cannot be honored. If the patient is committable, he can be involuntarily admitted. If not committable, he cannot be admitted. Nor could the agent go to court, because even if the court ruled the patient incompetent and made the agent the guardian, the guardian cannot sign in his ward unless the ward meets commitment criteria.

So what have we gained in Massachusetts with health care proxies for psychiatric admission? Virtually nothing.

**Treatment with medication**. Treatment with antipsychotic medication in Massachusetts with a health care proxy is even more convoluted than the admission decision tree. This is true, in part, because Massachusetts maintains a presumption of competency to refuse antipsychotic medication until ruled incompetent to do so by the Probate Court, and then provides for court authorization of treatment on a substituted judgment basis (Hoge, Appelbaum, & Geller, 1989). The various possibilities under Massachusetts' current psychiatric practice patterns, as declared by statutes and court decisions, and relevant to treating a person with antipsychotic medication who has a health care proxy expressing the principal's wishes about these medications, are shown in Table 1. Table 1 indicates the main value of the proxy would appear to be a clear statement of the patient's competent wishes about antipsychotic medication so that if he refuses treatment, and if the treating physician petitions the court for the authority to override the refusal, the court would be informed of the patient's

TABLE 1. Massachusetts Psychiatric Health Care Proxy and Antipsychotic Medication

| Proxy instructions | Agent's decision | Principal's action after determination of incapacity | Court review | Likely outcome |
|---|---|---|---|---|
| Accept Antipsychotics | Accept | Accept | None likely | Meds administered |
| Accept Antipsychotics | Accept | Refuses, overriding agent | Possible review on capacity | If no court review, no meds; if court review, likely meds administered per proxy based on substituted judgment standard |
| Accept Antipsychotics | Refuse | Refuse | Possible court challenge | Meds not administered absent court order based on substituted judgment standard |
| Accept Antipsychotics | Refuse | Accepts, overriding agent | None likely | Meds administered |
| Refuse Antipsychotics | Refuse | Refuse | None (except challenge to validity of proxy) | Meds not administered (unless proxy invalid and substituted judgment order by court) |
| Refuse Antipsychotics | Refuse | Accepts, overriding agent | None (except possible court review by agent) | Meds administered absent court order not to do so |

| | | | | |
|---|---|---|---|---|
| Refuse Antipsychotics | Accept | Accept | None likely | Meds administered |
| Refuse Antipsychotics | Accept | Refuses, overriding agent | Possible review on capacity | Meds not administered absent court order |
| No Specific Instructions | Accept | Accept | None likely | Meds administered |
| No Specific Instructions | Accept | Refuses, overriding agent | Court may be asked to determine capacity | If incapable, likely meds administered as per proxy |
| No Specific Instructions | Refuse | Accepts, overriding agent | Possible court review on capacity | If no court review, meds administered; if court review, depends on facts |
| No Specific Instructions | Refuse | Refuse | None (unless validity of proxy or conduct of agent is challenged) | Meds not administered (unless proxy is invalid or agent is overridden and substituted judgment order issued) |

wishes for treatment when competent, thereby informing itself of the basis for a substituted judgment decision. That is, rather than having to ascertain (by what means is enigmatic at best) what the patient would have decided about medication if competent, the court would have in front of it explicit directives on the matter of the patient's wishes about antipsychotic medication, articulated when the patient was competent.

## MASSACHUSETTS HEALTH CARE PROXIES IN PRACTICE

Are health care proxies being used for psychiatric patients? There are no systemic data to answer this question. In fact, on the National Association of State Mental Health Program Directors' (NASMPD) state policies database, only one state (Oklahoma) reported that the state mental health authority collected data on patients who have advance directives on file. To ascertain whether and how psychiatric advance directives are used for individuals with chronic mental illness who are in a state mental health facility, I examined this issue in one Massachusetts state hospital.

Why choose to look at psychiatric health care proxies in a state hospital? Many outpatient settings providing psychiatric services to those with serious mental illness are not covered under the PSDA. In informal discussions with executive and medical directors of community mental health centers in Massachusetts, it became clear that health care proxies are not routinely used. While acute inpatient settings are required to comply with the PSDA, how meaningful are these documents when mean lengths of stay are less than a week. This concern was confirmed when, at one university hospital, the nurse manager told me that health care proxies are systematically executed at the time of admission (overseen by the admitting nurse) while the medical director of the same unit told me proxies were not obtained (although he thought they should be).

The state hospital not only has a population that might maximally benefit from proxies, but state hospitals often do a better job of attending to the rights of the chronically mentally ill population than do other settings with long-stay populations (Barton, Mallik, Orr, & Janofsky, 1996). The state hospital might well be the best that is happening in terms of health care proxies for seriously, chronically mentally ill citizens. Finally, in a state that has gone to great lengths to keep people out of state hospitals (Geller, 1991), believes that almost everyone deserves community based services (Geller, Fisher, Simon, & Wirth-Cauchon, 1990) and creates those residential services (Geller & Fisher, 1993), the long-stay state hospital population is not terribly different from the long-stay community population. Examining the issues of health care proxies in the former population should give us insight into the issues we might face with the latter population as psychiatric health care proxies move from institutions to communities.

Worcester State Hospital (WSH) has a policy and procedure called "healthcare proxy." The policy statement reads, "In order to promote self-determination, it is the policy of WSH to inform all patients of their right under law to make informed health care decisions and to execute health care proxies. This information will be provided to patients at the time of admission, annually, and at other appropriate times." The procedure is spelled out in 11 steps, and includes statements that a copy of the health care proxy will be included in the transfer packet if a patient needs to go to a general medical hospital and will be included in the aftercare packet upon discharge. Attempts have been made, therefore, by WSH to provide all patients with the information and the process to effectively execute health care proxies, and even to assure continuity of the proxy when the patient leaves WSH. Does WSH meet the requirements of its own policy and procedure? And does it matter?

A point in time survey of WSH indicated there was a census of 161 patients, all of whom were nonacute patients since WSH only takes transfers from other inpatient settings. The demographic characteristics of the 161 patients were as follows: 116 males, 45 females; age range 19–85 years old, with a mean of 46.2 years old; 25 patients were under 30 years old and 13 were over 70 years old; 81 (50.3%) were voluntary patients, 65 (40.4%) were civilly committed patients, and 15 (9.3%) were criminally committed patients. The patients' length of stay ranged from 17 days to 14,576 days (almost 40 years) with a mean length of stay of 4.5 years and a median length of stay of 2.35 years. Diagnostically, 65 patients (40%) had schizophrenia, 7 (4%) had other psychotic disorders, 41 (25%) had schizoaffective disorder, 25 (16%) had bipolar affective disorder, 7 (4%) had dementia, 6 (4%) had organic personality disorder and 10 (6%) had other psychiatric disorders.

As indicated, for this population, health care proxies in relationship to psychiatric care and treatment are probably most important in terms of psychotropic medication and hospital admission. Of the 161 patients (see Figure 1), 71 (44%) had full guardians; this means they were judicially determined to be incompetent and would not be eligible to execute a health care proxy. Of the remaining 90 patients, 53 (33% of the population) each had a proxy in his/her medical chart. Thirty-four patients (21%) refused to sign them and 19 (12%) signed them.

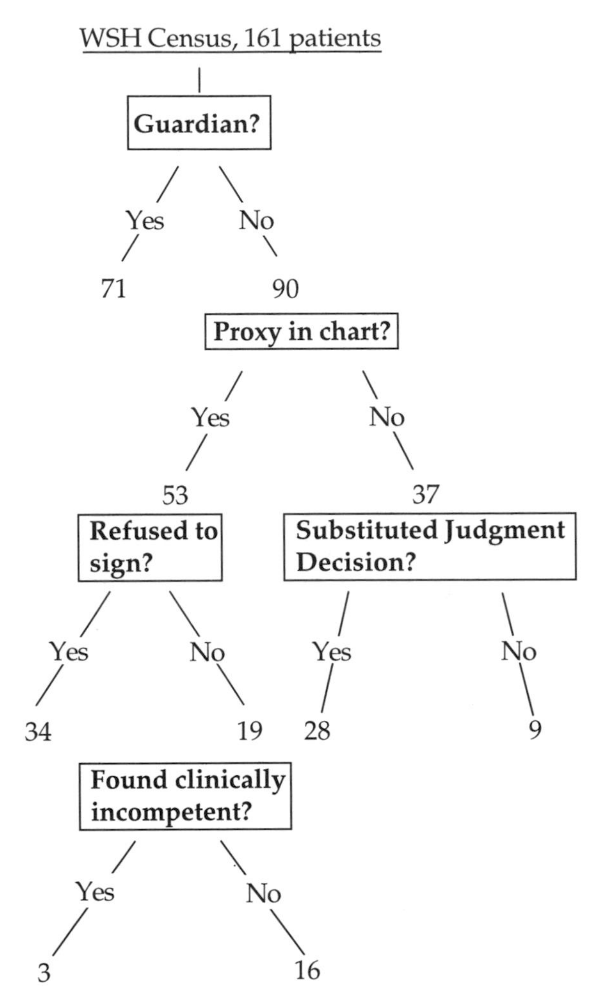


FIGURE 1. Health care proxies and psychotropic medication at Central State Hospital, Massachusetts.

Of those who signed, 3 were deemed clinically incompetent. Therefore, 16 of 161 patients, or 10%, had meaningful health care proxies in their medical charts.

Of the 37 patients whose charts had no proxy, 28 had substituted judgment decisions for psychotropic medication already determined by a court, so they had no ability to execute a meaningful proxy concerning medication. Therefore, of the 161 patients, no more than 9

(6%) were missing proxies that could have affected their use of psychotropic medication.

If we examine hospital admissions or legal status, i.e., the ability to indicate a preference about being hospitalized, in addition to the 16 patients with meaningful proxies, the salient question is how many of the 37 patients without proxies were not committed and could impact their admission status through a proxy? Eight of the 37 patients were civilly committed, five patients were criminally committed, and 24 patients were voluntary. It is not possible to know how many of the 16 patients with meaningful proxies and the 24 voluntary patients without them would have met commitment criteria even if they had had an agent authorized to refuse voluntary admission status. Therefore, of 161 patients, a maximum of 40 (25%) were even in a position to potentially have a health care proxy that could have meaningfully affected their inpatient status.

## DISCUSSION

The emerging literature on psychiatric advance directives or health care proxies proclaims many benefits to be accrued to patients by executing these documents. Benefits include: a more collaborative process between patient and provider; alleviation of tension among family members; facilitation of acceptance into community programs; opportunities for patients to make known their medication choices; specification of preferences in many areas such as which hospital and which providers to use, authority to contact providers who can visit when person is hospitalized, what child care arrangement should be made, type of activity therapy, and preferences for alternatives to hospitals; reduction in time involved in forced treatment; counteracting physicians' inappropriately dosing medication; prevention of physicians' use of medication as punishment; and the making of a "philosophy of life statement" (Rosenson et al., 1991; Rogers & Centifanti, 1991; Sales, 1993; Perling, 1993; Sherman, 1995).

One benefit not mentioned, but that might be added, is the clear opportunity for an individual to specify his "restrictiveness hierarchy." The oft referred to concept of least restrictive alternative is ill-defined, unmeasured, and treated as if it had universal applicability (Munetz, 1993). But restrictiveness is really a matter of individual experience. For one patient chemical restraint, and for another patient mechanical restraint may be the more restrictive. The advance directive would allow the patient to indicate what method of restraint should be used first, if restraint was to become a necessary intervention.

The list of purported benefits is all well and good. However, these benefits, as evidenced by the Massachusetts experience, have so far risen little above intentions. If the process of establishing advance directives for mental health services helps patients feel they have made "authentically informed choices" (Backlar, 1995), advance directives may have value. However, these documents also have the possible effect of disempowering patients, by there very existence causing a prohibition against questionably incompetent patients routinely assenting to treatment (Sales, 1993).

The value of psychiatric advance directives awaits empirical evaluation. These must be done from two points of view. First, does the process of being given the opportunity to execute a health care proxy have benefits in increasing patients' sense of participation in their overall treatment, whether or not these are subsequently deemed meaningful documents? If so, is the regularly scheduled review and modification of these documents, i.e., annual reviews, also useful? Second, do competently executed health care proxies in psychiatric settings have real meaning? Will they be routinely overridden by providers using legal avenues to address "bad decisions?" Or will patients withdraw or modify the proxies when changes in their mental status occur?

Both of these analyses need to be done in different psychiatric settings—long term inpatient, acute inpatient, outpatient—and with different demographic and diagnostic group to see if these variables have an influence on the outcomes. Further, if advance directives appear useful for some populations, can others be trained to make better use of health care proxies? There are preliminary data that education concerning health care proxies is needed by all involved in this process—patients, families and providers (Backlar & McFarland, 1996).

Psychiatric advance directives executed as health care proxies require outcome studies to better determine their effect and efficacy. Psychiatry does not need another intervention in the name of increasing patients' autonomy with no verifiable measures that the intervention actually accomplishes this end.

## REFERENCES

American Psychiatric Association Division of Government Relations (1992). State Update, December.

Appelbaum, P. S. (1991). Advance directives for psychiatric treatment. *Hospital and Community Psychiatry*, 42, 983–984.

Backlar, P. (1995). The longing for order: Oregon's medical advance directive for mental health treatment. *Community Mental Health Journal, 31,* 103–108.

Backlar, P., & McFarland, B. H. (1996). A survey on use of advance directives for mental health treatment in Oregon. *Psychiatric Services, 47,* 1387–1389.

Barton, C. D., Mallik, H. S., Orr, W. B., & Janofsky, J. S. (1996). Clinicians' judgment of capacity of nursing home patients to give informed consent. *Psychiatric Services, 47,* 956–960.

Cross, J., Fleischner, R., & Elder, J. (1994). *Guardianship and Conservatorship in Massachusetts*. New Hampshire: Butterworth Legal Publishers.

*Cruzan v. Director*. (1990). Missouri Department of Health, 110 S. Ct. 2841.

Geller, J. L. (1991). "Anyplace but the state hospital": Examining assumptions about the benefits of admission diversion. *Hospital and Community Psychiatry, 42,* 145–151.

Geller, J. L., & Fisher, W. H. (1993). The linear continuum of transitional residences: Debunking the myth. *American Journal of Psychiatry, 150,* 1070–1076.

Geller, J. L., Fisher, W. H., & Simon, L. J., Wirth-Cauchon J. (1990). Second-generation deinstitutionalization, II: The impact of *Brewster v. Dukakis* on correlates of community and hospital utilization. *American Journal of Psychiatry, 147,* 988–993.

Hoge, S. K., Appelbaum, P. S., & Geller, J. L. (1989). Involuntary treatment. In A. Tassman, R. E. Hales, & A. J. Frances, (Eds.) *American psychiatric press review of psychiatry* (pp. 432–450). District of Columbia: American Psychiatric Press.

Lennon, J. (1980, 1981). *Beautiful Boy*. Lenono Music.

Matter of Rosa M. (1991). 597 N.Y.S. 2d 544.

Munetz, M. R., & Geller, J. L. (1993). The least restrictive alternative in the postinstitutional era. *Hospital and Community Psychiatry, 44,* 967–973.

Perling, L. J. (1993). Health care advance directives: Implications for Florida mental health patients. *University of Miami Law Review, 48,* 193–228.

Rogers, J. A., & Centifanti, J. B. (1991). Beyond "self-paternalism": Response to Rosenson and Kasten. *Schizophrenia Bulletin, 17,* 9–14.

Rosenson, M. K., & Kasten, A. M. (1991). Another view of autonomy: Arranging for consent in advance. *Schizophrenia Bulletin, 17,* 1–14.

Sales, G. N. (1993). The health care proxy for mental illness: Can it work and should we want it to? *Bulletin of the American Academy of Psychiatry and the Law, 21,* 161–179.

Sherman, P. S. (1995). Advance directives for involuntary psychiatric care. *Proceedings of the National Symposium on Involuntary Interventions: The Call for a National Legal and Medical Response*. Texas, University of Texas.

Szasz, T. S. (1982). The psychiatric will. *American Psychologist, 37,* 762–770.

Part VI

# RESEARCH ETHICS

CHAPTER 12

# ETHICS IN NEUROBIOLOGICAL RESEARCH: ONE CONSUMER/PROVIDER'S PERSPECTIVE

FREDERICK J. FRESE III

The National Bioethics Advisory Commission (NBAC), is an eighteen member, presidentially appointed commission charged to focus on the protection of the rights and welfare of human research subjects. In November of 1998, as NBAC was completing its report and recommendations on *Research Involving Persons with Mental Disorders That May Affect Decisionmaking Capacity* (1998), numerous accusations were made in the press concerning the conduct of psychiatric research in this country.

A *Washington Post* article on November 16, 1998 (Weiss, 1998a) described the mentally ill as persons who can neither consent or effectively refuse, to participate in research. It then quoted Beverly Post, who complained about her son's treatment in a research study at the Maryland Psychiatric Research Center, saying the doctors gave her son amphetamines to exacerbate his symptoms for study. She claimed two years of his life were lost there. In response to this and similar testimony the *Post* reported on November 18, 1998 (Weiss, 1998b) that NBAC would make numerous recommendations for change. The article referred to the checkered history in which mentally ill people have been the unwitting victims of ethically dubious experiments.

While the *Washington Post*'s coverage may have caused some concern in the public's mind about what may have been transpiring in the nation's psychiatric research centers, the coverage in the *Boston Globe* concerning this issue was much more detailed and much more damning. In a four part series beginning November 15, 1998, the *Globe* (Witaker & Kong, 1998) detailed the plight of Shalmah Prince, who was given apomorphine, in order for a researcher to find out if it could provoke psychosis. After the injection she had to be placed in leather restraints, and descended into madness that did not fully subside for 10 days. The *Globe* series revealed that there were numerous studies where mentally ill patients were injected with drugs such as L-dopa, tetrahydrocannabinol, and ketamine. These studies were designed to intentionally exacerbate delusions and hallucinations in over 1200 schizophrenia patients. There were also studies where they deliberately stopped giving medication to stabilized schizophrenic patients to see how quickly they became sick again. The authors of the articles pointedly reminded readers of the infamous Tuskegee syphilis studies where researchers denied treatment to poor infected black men in the rural South for some forty years until 1972, and the similarly embarrassing Cold War radiation studies.

The *Globe* pointed out that the federal Office of Protection from Research Risks (OPRR) had found fault with researchers at the University of Maryland, the University of California, Los Angeles, and the National Institute of Mental Health, and had ongoing investigations of three other research facilities. It went on to quote numerous well known and respected researchers who acknowledged conducting challenge and washout studies and attempted to justify their actions. It stated that in a three month investigation the paper had "found a trail of both harm and deceit."

In ensuing installments *Globe* reporters detailed numerous incidents involving psychiatric research where medical ethicists had raised serious ethical concerns. The newspaper described schizophrenia research as being a "landscape tarnished by the greed of some rogue investigators and repeated incidents of patients being harmed." The authors detailed researchers' shortcomings under bold print headings such as: "Researchers seduced by lure of lucrative rewards"; "Researcher caught falsifying records"; and "Messy picture often buried in a marketing glow."

The final installment of the *Globe*'s revealing series reported that NBAC called for tougher research standards to protect the mentally ill, including "boosting representation of people with mental disorders on local review boards and requiring greater researcher accountability."

In light of the *Post*'s and *Globe*'s revelations, what course of action, one might ask, is in the best interest of persons with serious mental illness?

## A CHANGED WORLD

To answer this question about the best interests concerning the involvement of mentally ill persons as subjects in research, one might suggest that we solicit the opinion of these persons themselves. Or if for some reason that course of action is not practical, then perhaps we could solicit the opinions of their family members or perhaps of their trusted friends. Until a short time ago such an approach however would have seemed impractical if not ridiculous. This is because until recently, persons with serious mental illness generally did not recover to the point where their input would have value. And those who did so recover, would almost never identify themselves as persons in recovery from schizophrenia or any of the other serious mental illnesses. The shame and stigma accompanying these disorders was so great that, also, very few relatives of these persons were willing to let it be known that there was insanity in the family. But this situation has changed markedly during the past generation, and particularly since the founding of the National Alliance for the Mentally Ill (NAMI) two decades ago. The rise and complexity of the consumer and family advocacy movement has been chronicled elsewhere (Frese, 1998), but clearly the inclusion of the voices of those who have been personally touched by these disorders has implications of major proportion for the processes by which decisions are made concerning research on (and for) these individuals.

Now that persons who have been diagnosed with these disorders and their family members are increasingly willing to identify themselves and are often to be seen crying out to have their voices heard and their views respected, there is growing realization that consumers and family members should be afforded a place at the decision making table. I am suggesting, however, that the process of including consumers and family members in such decision making should not be limited to their involvement in individual research studies. In order to provide dignity, fairness, and a modicum of justice to those who allegedly have been victimized by the research establishment in the past, efforts must be made to include consumers and family advocates at all levels of debates concerning these issues.

Until very recently researchers might listen to attorneys, scholars, ethicists, and others who purported to represent the mentally ill, but input from the victims of these disorders was not actively elicited. Indeed, in many quarters today, papers are still written, presentations delivered, and conclusions drawn concerning research activities with no acknowledgment of the developing body of opinion emanating from those with most at stake in this process.

Because I am a person with mental illness, I am finding that I am increasingly distrustful of academicians, and other "experts" who write papers or otherwise issue opinions without any reference to the increasingly available body of consumer and family literature on this issue. In examining recent published opinion however, I find two examples of broad discussion of psychiatric research where coverage of ethical issues gave full exposure to both the consumers and family members, as well as to the more traditional players in this arena.

While, as I mentioned, there have been numerous isolated incidents of questionable research behavior exposed and discussed in the past, the most extensive forum with the broadest participation for exchange of views on these issues, arguably occurred in 1994. In that year an entire issue of the *JOURNAL of the California Alliance for the Mentally Ill* (*CAMI JOURNAL*) was published focusing on this topic. This particular issue was entitled, *Ethics in Neurobiological Research with Human Subjects*, and consisted of some 29 articles contributed by eminent researchers and research administrators, biomedical ethicists, legal scholars, and the like, as well as former patients, family members and like-minded activists. Many of these articles pointed out areas of difficulty in psychiatric research. Others primarily defended the status quo. Numerous recommendations for improvement were suggested.

Another broad based forum for interaction among such diverse contributors concerning these issues occurred in Baltimore Maryland, January 7–9, 1995. This was the convening of the First National Conference on Ethics in Neurobiological Research with Human Subjects. Over 150 of highly respected scholars, ethicists, researchers, federal administrators, and advocates, as well as consumer and family activists gathered for three days of consecutive presentations, many of which generated pointed questions and heated discussion of the information and opinions presented. An overview of this conference was published in 1997 in a volume, entitled *Ethics in Neurobiological Research with Human Subjects*, which contained a conference summary, seven background papers, and 26 papers (many of them subsequently revised) that had been presented at the conference.

In the following portion of this chapter I will review the opinions of many of those who shared their views during these two important developments. Although I provide brief snapshots of these authors' perspectives on this volatile issue, it is hoped they give a fair exposure to each of these authors' perspectives. Subsequently the readers' attention will be drawn to what I feel is a concise and parsimonious approach to thinking about these issues. As a contributor to the research ethics issue of the *CAMI JOURNAL*, as well as being a participant at the Baltimore Conference, I do not claim detached objectivity concerning what transpired during these events. On the other hand, having been personally involved in the discussions during the Baltimore Conference, which was also attended by many of the other contributors to the *CAMI JOURNAL* issue, I may have gained additional insights from observing the "demeanor of the witnesses" and other parties involved in this overall process.

## THE CAMI JOURNAL: ETHICS IN NEUROBIOLOGICAL RESEARCH WITH HUMAN SUBJECTS

In the early 1990s, Greg Aller, a young man with schizophrenia, had been a research subject in a UCLA/NIMH-funded protocol that focused on relapse. His father, Robert Aller, an active member in the California Alliance for the Mentally Ill (CAMI), watched as his son decompensated and began behaving in threatening, disorganized manner. When Mr. Aller attempted to bring his son's plight to the attention of the authorities at UCLA, he was not satisfied with the action—or as he saw it—the inaction of the university officials. Mr. Aller had considerable experience with the media, thus he was able to bring national attention to the plight of his son and the other individuals who had unfortunate experiences as a result of their being research subjects in the UCLA project. Because of the subsequent media attention, including extensive coverage in such media heavyweights as, *Time Magazine, The New York Times*, and *The Larry King Show*, what transpired at UCLA began to raise national consciousness concerning the propriety of how research was being conducted on persons with serious mental illness throughout the country.

Dan Weisburd, a former CAMI president, and editor of the widely distributed, *CAMI JOURNAL*, happened to have a long-time friendship with the Aller family. He decided to put together an issue of his journal, inviting many respected professionals and advocates to express their opinions concerning "Ethics in Neurobiological Research", which he used as the title of that quarterly issue, published in 1994.

According to Mr. Weisburd (1994), Mr. Aller and his son eventually declined to submit an article. Therefore, the article submitted by a psychologist from the UCLA researchers responsible for the study was not published because the editor felt he wanted to keep a sense of balance. Nevertheless, the articles that were published blended to produce an intriguing forum on the many very different views concerning ethics in psychiatric research. In some ways this *CAMI JOURNAL* issue can be seen as type of a verbal sporting event. On one side are those attacking the research establishment, accusing it of all sorts of reckless, even diabolical, activities. On the other side are those who defend the way research is conducted, arguing that no progress is made without research, and that the mishaps that do occur are very rare. The defenders typically called for increased partnership, collaboration, and continual need to balance the risks and benefits involved.

The accusers in this drama tended to be consumers and their family members, along with friendly scholars, ethicists, lawyers and other advocates. Some of their criticism was rather disturbing. For example, Vera Hassner (1994), who was identified as serving on the Board of Directors of AMI of New York State, argued that "a fundamental ethical line had been crossed" when individuals who lack capacity to understand risks are recruited into invasive experiments without any expected benefit for themselves. She claimed that repeated drug wash-outs, schizophrenia relapse studies, and the like are "more torture than science." She questioned whether families should be endorsing the "sacrificing of other people's children for the sake of scientific advancement."

George Annas (1994), a professor of health law at Boston University, weighed in by recalling the atrocities committed by the Nazi physicians during World War II and reminding readers that the Nuremberg Code was produced as a result of the trials of those physicians and other Nazi war criminals. He stressed that the requirement for informed consent on the part of research subjects requires that such consent be competent, voluntary, informed, and comprehending. He then related the tale of a serviceman who had been secretly given LSD in 1958 as part of an experiment to determine its effects. He went on to tell of radiation experiments conducted by the Atomic Energy Commission which he said violated the Nuremberg Code. He said that proposals had been made to Congress to compensate the victims. He raised the specter of numerous other recent medical experimentation scandals. These included a 1963 study in Brooklyn, N.Y., where terminally ill cancer patients were unknowingly injected with live cancer cells in order to test their immune response and, of course the famous "Tuskegee Study," where for over three decades effective treatment was kept from a group of syphilitic black males so that the natural course of the disease could be studied. He also mentioned the notorious "Willowbrook experiments," where retarded children were deliberately infected with hepatitis. He cited the landmark article by Professor Beecher (1966) of the Harvard Medical School, Ethics and Clinical Research, which revealed some 22 American unethical experiments that had occurred after the promulgation of the Nuremberg Code. He commented on glitzy experimentation and ambitious physicians, and how ethics and law may take a "back seat" when fame and potential fortune for researchers is involved.

Adil Shamoo (1994), a professor of Biological Chemistry, at the School of Medicine, the University of Maryland at Baltimore, wrote an article that addressed washout and relapse studies. Professor Shamoo reviewed over forty studies where relapse was either part of the design or an expected consequence. He found that over 940 of the involved research subjects had relapsed, and that some had committed suicide. He decried the community of scholars for failing to speak out about these abuses. He made several specific recommendations for change in regulations, including that "All research involving persons with mental illness should have direct or potential benefit for that individual patient."

Janice Becker (1994) is a State of Maryland AMI Board member and mother of a daughter who spent thirteen years as an inpatient in mental hospitals. Her article relates the trauma endured by her daughter when she was experiencing repeated washouts as a subject in a research protocol. She also told of the many promises that she claimed were broken by those in charge during this time. She questioned the value of such research and particularly the propriety of washouts studies.

Leonard Rubenstein (1994), then the Director of the Bazelon Center for Mental Health Law, cast particularly effective accusations against psychiatric researchers. He cited activities of Ewen Cameron, one of the most prominent psychiatrists of his time, and his now infamous "psychic driving" and "depatterning" experiments during the 1950s. Dr. Cameron was president of the American Psychiatric Association, and had been a psychiatric examiner of Nazi War criminals. Nevertheless, this distinguished psychiatrist used intensive electroshock, isolation of patients, and put patients to sleep for up to ten days in order to "disinhibit their natural defenses." He also injected patients with LSD, curare, and addictive drugs as part of his experiments. Patients often did not know they were part of these experiments and were ignored when they asked for these actions to stop. Rubenstein particularly laments that, although all these activities were published in major journals and presented before sophisticated professional audiences, none of these members of the research community complained about informed consent, possible harm to patients, etc. Based primarily on Dr. Cameron's research ventures, Mr. Rubenstein concluded that "we cannot leave it to researchers to balance what they may see as competing values" and that "rules must be established and there must be a mechanism to see to their adherence."

Martin L. Smith (1994) is an associate staff member in the Department of Bioethics at the Cleveland Clinic Foundation, a national referral center. Mr. Smith's article also lamented that "the history of human research is checkered with abuses and exploitation." His strong recommendation was that consumers need to become their own advocates in order to safeguard their interests and rights, and pay careful attention to the consent form.

Bonnie Shell (1994) described herself as a person with bipolar disorder, dedicated to speaking out for other "labeled people" when they are ignored. Her article, too, decried the CIA-sponsored activities of Dr. Ewen Cameron, but also castigated the CIA for much other questionable research including the use of electromagnetic fields, brain washing, high intensity strobe lights, and other "non-lethal weapons" on unsuspecting victims of research.

Dianne N. Irving (1994) a former research biochemist at NIH, contributed an article in which, after referencing the Nazi atrocities, claimed that "if psychiatric research is to be done, it should be therapeutic research only—done only for the direct benefit of that patient" and that "no persons should feel pressured into taking part in an experimental research protocol for someone else's good." She also lamented the effects of L-dopa and amphetamine challenge studies on human research subjects. Irving believes these experiments intentionally provoked devastating relapses that happened to patients once they returned to the community where they could not be effectively monitored. She claims that it is impossible for incompetent patients to have an appropriate surrogate represent their interests during experiments because such patients could not have been competent to chose such surrogates. Furthermore, she suggested that conflicts of interest are a serious problem of researchers and that "no one seems to be accountable to anyone for anything."

The other authors, who were researchers themselves, understandably took a somewhat different posture when they presented their view of psychiatric research activities.

Jack Barchas, M.D., in an article co-authored by Isaac D. Barchas (1994), started out by comparing psychiatric treatment today with how it had been forty years ago. He contrasted patients being wrapped in ice-cold sheets and being given continual electroconvulsive therapy with the greatly improved conditions today where 70–80% of patients with manic depressive disorder are successfully treated. The authors claimed that research changed these conditions for the better. They further pointed out that in their opinion the status quo is not good enough and research must proceed in order to bring about further improvements. They strongly argued for participation of subjects in research projects even when they are only motivated by altruism.

John Kane (1994), Professor of Psychiatry at Albert Einstein College of Medicine in Bronx, another renowned psychiatric researcher, also strongly emphasized the need for balance in his message. He addressed the necessity for conducting placebo-controlled trials, and defended clinical trials in which people are allowed to become ill again with the proviso that proper monitoring mechanisms are in place.

Ram Kaminski (1994), a researcher with the Department of Psychiatry at Mount Sinai School of Medicine, made a plea for establishing a trusting and true partnership with his subjects. He perceives that "most patients on most days [are] capable of understanding" the protocols, "provided everything is spelled out in a clear manner." Dr. Kaminski did not express particular concern regarding the fact that many patients may be incapable of understanding the risks and benefits of being a research subject.

William Wirshing (1994), an associate professor of Psychiatry at UCLA, in a short article, also stressed the importance of balancing risks and benefits. He claimed that "most clinical experimentation takes place in a setting that is safer, better funded, and far more enlightened and informed than non research medical treatment." Interestingly, he was most outspoken when lamenting how "complex, time consuming, and highly scrutinized informed consent procedures have become." He claims that "the process has become so legalistically elaborate that it is difficult if not impossible to properly inform a psychiatrically ill patient." Dr. Wirshing only gave the briefest allusion to adverse consequences of research. He referred to "the every so often rare tragic example" of adverse consequences, and stated that a typical image of experimenter malice "is far more a theatrical illusion than a medical reality."

Lewis Opler (1994), a clinical professor of Psychiatry at Columbia University College of Physicians and Surgeons, argued for more research to be done in clinical settings as opposed to dedicated research units, which he suggests may be geared too much toward the generation of publications and the chasing of research grants, while being insensitive to the real issues of consumers. Dr. Opler did not directly address issues of informed consent and challenge studies that occupied the attention of so many of the other *CAMI JOURNAL* contributors.

Jeffrey Lieberman, then Director of Research at Hillside Hospital, Long Island Jewish Medical Center and his co-author, Judith Sloan (Lieberman & Sloan, 1994), argued that medical science is currently on the brink of major breakthroughs that can only come about through research. These authors did acknowledge scientific misconduct, fraud, and patient abuses that had recently been reported and condemned them as reprehensible. Additionally, they argued for proper representation on Institutional Review Boards (IRBs) that could include patient representatives and also called for special efforts to ensure proper implementation and compliance with rules and regulations.

Drs. Nassir Ghaemi and Edward Hundert (Ghaemi & Hundert, 1994), physicians associated with the Harvard Medical School, pointed out that several governmental bodies have been established which have had responsibilities for assuring that research be in done in an appropriate fashion. These include the National Commission for the Protection of Human Subjects of Biomedical and Behavioral Research, the President's Commission for the Study of Ethical Problems in Medicine and Biomedical and Behavioral Research, the Ethics Advisory Board, the Advisory Committee on Human Radiation Experiments, and NBAC. In response to the question, "Does the medical community accept ethical guidelines and put them into practice?" The authors answered "usually" and "most of the time." However, the authors went on to suggest that most scientific misconduct takes the form of "fudging the data" as opposed to mistreatment of the research subjects. Additionally, these authors also claimed that patients have a "partial obligation to engage in medical research," because they not only have a duty to themselves "but [also] to other human beings who might have or might develop the same illness." These authors did, however, allude to Nazi atrocities, Beecher's "Ethics in Human Experimentation" article, and the Tuskegee research, saying that patients, families and friends should recognize that baser motivations, such as desires for fame, money, and status, sometime spur research activity and that they should "never confer carte blanch trust upon medical researchers." While Gaemi and Hundert must be complimented for their forthrightness on this issue, Burt Angrist (1994), a professor of Psychiatry at the NYU School of Medicine, should be noted for being particularly courageous in his article. According to Ms. Vera Hassner (1994), many researchers who conducted induced psychotic relapse experiments had been invited to submit articles for this *CAMI JOURNAL* issue, but Dr. Angrist was the only one who accepted the invitation. In his article he allowed that such experiments may "now seem barbaric and dangerousness." However, he did an excellent job of laying out the background, rationale, potential advances, and benefits of his research where he had administered amphetamine to psychiatric patients in order to provoke psychotic symptoms. Although he made a very good case that his research was ethical and worthwhile, he also stated that he does not think he would repeat the studies now, due to redundancy and limitations on the potential advances to be gained.

In other articles, Drs. Frederick Goodwin (1994), David Shore and Kate Berg (Shore, Berg, & Mullican, 1994), and others who are, or have been, officials with the National Institute of Mental Health reviewed and commented on many of the rules and procedures governing neurobiological research and Dr. Richard Meibach (Zarin & West, 1994), of the Janssen Research Foundation, reviewed the process by which new drugs are approved. These articles were helpful in giving the reader a better understanding of the complexities that govern the system under which research is conducted and new treatments are approved. Several other authors, including Deborah Zarin (Shamoo, 1997), with the American Psychiatric Association, contributed articles which tended to be more peripherally related to the issues being raised in this debate.

Overall, the *CAMI JOURNAL* issue highlighted general agreement among the contributors that research needs to go forward, and although it is the researchers who must have primary responsibility for planning and conducting such research, most parties agreed that efforts should be made to increase the involvement of consumers, their family members, and other advocates in the process of planning and monitoring these endeavors.

As a consumer/professional/advocate reviewing these articles I am struck by the overall consensus that attempts should be made to ensure that subjects maximally understand and agree to the purposes and procedures of the research. Repeatedly, the authors call for partnership between the researchers and the research subjects, or their trusted representatives. It would appear to me that the most important aspect of this issue focuses on determining the degree of competency of subjects involved and the process of determining who should be their trusted representative or assistant in the event they are determined to have diminished capacity for understanding and agreeing to become subjects in research endeavors.

## THE BALTIMORE CONFERENCE ON ETHICS

Subsequent to the publication of the *CAMI JOURNAL* issue, one of the contributors, Dr. Adil Shamoo, of the University of Maryland at Baltimore convened a three day conference, inviting many of those who had written for the *CAMI JOURNAL*, as well as numerous other distinguished interested figures, to participate. The convening of this meeting involved considerable acrimony. Dr. Shamoo (Shamoo, 1997) had made a presentation at the 33rd annual meeting of the American College of Neuropsychopharmacolgy (ACNP) in December, 1994, where he took exception to certain activities of the psychiatric research establishment. He mentioned that due to criticism by the media and the general public, the perception of psychiatric research was becoming one "that approaches Nazism." Apparently these statements, presumably along with other factors, initially precipitated the members of ACNP, by way of a letter from its president, Dr. David Kupfer, to call for a boycott by those ACNP members who were scheduled as speakers at the conference. After some intense negotiations, all these speakers participated, but there remained a certain degree of discordance among the participants as the conference came together.

An overview of the Baltimore Conference has been provided by Shamoo and Cassidy (Shamoo & Cassidy, 1997). About half those making presentations were authors of articles in the *CAMI JOURNAL* issue on research ethics. Many of these amplified and intensified their previous positions. These amplifications along with the contributions of those who had not written for the *CAMI JOURNAL* research ethics edition, substantively broadened the scope of the overall discussion.

The Baltimore Conference included an increased number of consumer and family presenters. These included including Peggy Straw (1997), one of the founders of NAMI, who chastised researchers who conduct research from a "family blaming" perspective, and Jean Campbell (1997), a doctoral level consumer/researcher who stressed "the need to reach consensus on issues of research ethics in collaboration with human subjects with neurobiological disorders" (p. 303). Of particular interest was the presentation by Bob Aller and his son, Greg. They summarized their own experiences and those of three other families who had complaints against researchers at UCLA. Greg Aller graphically described his experience as a research subject, of being made to relapse into psychosis, where he threatened his mother with a knife in order to scare the Devil out of her, lapped water from a toilet like a dog, and plotted to assassinate the president with poison gas. The Allers detailed numerous complaints about officials who failed to respond to inquiries and even issued false information. And, they described the series of events that led up to their filing a formal complaint with the National Institute of Health's Office for Protection from Research Risks (OPRR).

Janice Becker (1997) elaborated on her testimony in her *CAMI JOURNAL* article, still more vehemently decrying her having been "lied to and used" (p. 185) during her daughter's dangerous ordeal at the hands of research staff at the Maryland Psychiatric Research Center. She also referred to three other families who had recently contacted her concerning their anger about their mistreatment by the same research group.

In addition to more inclusive testimony by consumers and family members, the conference offered researchers expanded opportunity to justify their activities. Dr. William Carpenter (1997), Director of the Maryland Psychiatric Research Center, objected vociferously to *ad hominem* attacks on individual researchers and decried what he saw as a lack of evidence supporting the accusations being leveled. He warned that such "uninformed and misdirected censure derails the process" of scientific investigation, and of the dangerousness of "an ambiance of accusation ... without documentation of legitimacy" (p. 226).

The UCLA Clinical Research Center (Statement, 1997) also submitted a paper. In that no staff members from that Center were active conference participants, this document was read to the audience after the Allers' presentation. The brief paper repeatedly referred to the accusers using terms such as: "misrepresenting the facts"; "false and unfounded accusations"; and "baseless speculation." On the positive side, the paper stressed that the UCLA Center was "striving to forge a successful alliance among those afflicted with mental illness, their clinicians, and their families" (p. 173).

Appearing with the above combatants, were several high ranking psychiatrists and others who had not been not contributors to the *CAMI JOURNAL* ethics issue. Dr. Herb Pardes (1997), former president of the American Psychiatric Association, commented on what he called the "acrimony that arose in connection with [the] conference," but praised the bringing of "people from the array of constituencies together" for discussion.

Drs. Sunderland and Dukoff (1997), from NIMH, recounted positive experiences they felt they had had employing a durable power of attorney (DPA) with cognitively impaired patients. Most of their comments were based on experiences with Alzheimer's Patients.

Dr. Jay Katz (1997), of the Yale Law School, who had chaired the sub-committee appointed to review the Tuskegee Syphilis Study, delivered the final paper of the Conference. He stressed the near-universal agreement reached by the participants on the necessity for continuing research. But he expressed concern for dignitary harm to subjects whose consent and participation are manipulated. Considering informed consent to be a cornerstone on which research morality must rest, he focused heavily on what should be revealed during this process, but he did not devote much attention to the process of determining the degree of competency of the patient-subjects.

## SEMINAL ISSUES

To this consumer/professional/advocate, a cornerstone of this controversial issue involves the degree of competency of the research subject. If the person with schizophrenia or other serious mental illness is reasonably competent, then other questions, such as the adequacy of the informed consent document, whether placebos should be employed, the propriety of challenge studies, etc., can then be addressed, because the person whose rights are in question would have the ability to consider and decide matters concerning these issues. But if these persons are clearly not competent, then questions about selecting possible surrogates, who then might be able to decide such issues for the patient/subject, come into play. Obviously then, pivotal issues in this debate concerns how we go about determining the degree of competency of the prospective subjects.

As I addressed in my article in the *CAMI JOURNAL* (Frese, 1994), I see great similarities between the process of determining competency (or capacity) regarding the right of patients to refuse medication and other forms of mandated treatment and making determinations of competency in research endeavors. From this perspective, the presentation at the Baltimore Conference by Dr. Paul Appelbaum (1997), of the Department of Psychiatry at the University of Massachusetts Medical Center, was particularly worthy of note. Dr. Appelbaum focused squarely on this issue, stressing the similarity of the question of competency in matters of research with the question of competency in matters of treatment; he pointed out that while there is little literature focusing on the former, there is a considerable body of literature regarding the latter. He argued that the standards developed for determining competency to consent to treatment can in large part be applied to determinations of competency in research. He went on to point out that four standards have been developed concerning treatment decisionmaking competence. These are:

1. the ability to communicate a choice;
2. the ability to understand relevant information;
3. the ability to appreciate the nature of the situation and its likely consequences; and
4. the ability to manipulate information rationally.

More importantly, Appelbaum points out that while schizophrenic patients may generally be much more impaired than most people or even other categories of mentally ill patients, there is a broad range of abilities within even this group. He also argues that the level of capacity required by participant subjects should be a function of the degree of risk involved in the project. Research projects involving little risk to the subjects should not require the same level of subject competence as projects where risks are substantially higher.

Appelbaum also addressed the difficulty of involving patients who have little decisionmaking capability as subjects in research. He suggests that it may be very difficult, but he does welcome suggestions concerning creative approaches to authorizing surrogate decision makers.

If one accepts the premise that determination of capacity to consent to treatment and capacity to consent to becoming a research subject are similar, one should then perhaps look more carefully at methods for determining capacity to consent to treatment. In a recent collection of articles concerning the possible value of mandated treatment (Munetz, 1997) several authors (Frese, 1997; Wilson & Wilson, 1997) refer to the use of a capacity assessment instrument that has been in use in the state of Ohio for over a decade now. This device (Morgan, 1987) has been found useful for determining the capacity of patients to make rational decisions concerning their opposition to accepting prescribed treatment for their conditions.

Although the Ohio instrument addresses the factors delineated by Appelbaum, it renders an either/or determination regarding capacity and does not address where a client may be functioning along what

Appelbaum sees as a broad range of competency, Munetz and Frese (2001), (Morgan, 1987) are also addressing the issue of degree of recovery and its importance in making determinations. These authors point out that ability of patients to make competent decisions related to their well-being is a function of their degree of recovery.

Another recommendation that I (Frese, 1997a) made at the Baltimore Conference was that a recovered person be involved in any review of decisions concerning patient involvement as subjects in research. This recommendation was described as "novel and worthy of consideration" (p. 15) in the conference summary (Shamoo & Cassidy, 1997) and is essentially the same recommendation being made by Munetz and Frese (2001) regarding the process of arriving at decisions concerning mandated treatment.

## CONCLUSIONS

The world of conducting research on persons with serious mental illness is changing dramatically with the involvement of recovering patients and their family members in the process. As the opinions vented in the *CAMI JOURNAL*, the Baltimore Conference, and the press reaction in the wake of the NBAC report reveal, psychiatric researchers should no longer continue to conduct "business as usual." No longer may they assume that just because they have made good progress, and feel what they have been doing is best for the seriously mentally ill, should they presume that they may continue "on their usual course." The patients that they have helped restore to competency, often along with their family members, are demanding that their voices also be heard when suggestions are discussed as to how conditions can be improved for research subjects and potential research subjects.

While the opinions of recovering persons, families, and sympathetic advocates during these forums may appear to have been acrimonious, the fact they occurred at all is a highly salutary sign that progress is being made. Acrimony engenders heat, and it is from such heat that light can be forthcoming.

Considering the forgoing brief overview of opinions voiced during the exchanges above, this observer makes the following observations/recommendations:

1. Consumers, individually and collectively, be given, as a matter of course, maximal input in all aspects of research that is done

where consumer/patients are to be the subjects. This should include, not just having input as to whether potential subjects have been afforded informed consent, but input concerning all phases in the design, implementation, monitoring and review of such projects. My suggestion to this effect at the conference entailed only that recovered patients be involved in any review of the informed consent mechanism, but upon reflection I think it better to try to include recovered persons in all phases of the research enterprise;

2. Due to an inherent inability of many consumers to process and communicate information in an effective manner, their "next of kin" and other caring family members, should be the first persons, after the consumers themselves, to be invited to assist by representing the interests of their family members;
3. As the shame and stigma of mental illness diminishes, increasingly we are finding that the researchers themselves are family members of persons with seriously mentally ill persons, e.g., see Pardes (Pardes, 1997), and upon occasion, they are themselves consumers. I would encourage more researchers to become open about their own personal and family experiences with these disorders, so that others might understand that there often already is consumer and especially family input into the process of designing and conducting research; and
4. The argument can be made that consumers, with their impaired cognitive abilities and relative lack of training, simply cannot understand the complex concepts and technical language used in research sufficiently for them to be able to make meaningful contributions. In response, I suggest that every effort should be made to allow recovering consumers, and their family members, to receive the type of training that would enable them to better make contributions in this regard. This recommendation not only includes welcoming consumers and family members at conferences and seminars, but also encourages them to attend professional and graduate schools where they can obtain the training necessary for them to enter the mental health and research professions.

Currently there is egregious discrimination on the part of health care professions against persons with mental illness. Such discrimination should be attacked and reversed. Only when the organized health-care professions begin treating mentally ill persons with dignity and respect in all aspects of their endeavors will we be able to effectively turn around our history of marginalizing the mentally ill, and reducing them to being—like the laboratory animals—primarily of value as powerless, potential subjects of scientific research.

## REFERENCES

Aller, R., & Aller, G. (1997). An institutional response to patient/family complaints. In A. E. Shamoo (Ed.), *Ethics in neurobiological research with human subjects: The Baltimore conference on ethics* (pp. 155–172). Amsterdam: Gordon & Breach.

Angrist, B. (1994). Ethical issues regarding prospective studies of amphetamine psychosis. *The Journal of the California Alliance for the Mentally Ill, 5*(1), 32–35.

Annas, G. J. (1994). Experimentation and research. *The Journal of the California Alliance for the Mentally Ill, 5*(1), 9–11.

Appelbaum, P. S. (1997). Patients' competence to consent to neurobiological research. In A. E. Shamoo (Ed.), *Ethics in Neurobiological Research with Human Subjects: The Baltimore Conference on Ethics* (pp. 253–264). Amsterdam: Gordon & Breach.

Barchas, J. D., & Barchas I. D. (1994). The imperative for research on severe mental disorders. *The Journal of the California Alliance for the Mentally Ill, 5*(1), 18–19.

Becker, J. (1997). Expanding a mother's testimony. In A. E. Shamoo (Ed.), *Ethics in neurobiological research with human subjects: The Baltimore conference on ethics* (pp. 183–186). Amsterdam: Gordon and Breach.

Becker, J. C. (1994). A mother's testimony. *The Journal of the California Alliance for the Mentally Ill, 5*(1), 17.

Beecher, H. (1966). Ethics in clinical research. *New England Journal of Medicine, 274*, 1354–1360.

Campbell, J. (1997). Reforming the IRB process: Towards new guidelines for quality and accountability in protecting human subjects. In A. E. Shamoo (Ed.), *Ethics in neurobiological research with human subjects: The Baltimore conference on ethics* (pp. 299–303). Amsterdam: Gordon and Breach.

Carpenter, W. T. (1997). Schizophrenia research: A challenge for constructive criticism. In A. E. Shamoo (Ed.), *Ethics in neurobiological research with human subjects: The Baltimore conference on ethics* (pp. 215–228). Amsterdam: Gordon and Breach.

Frese, F. (1994). Informed consent and the right to refuse or participate. *The Journal of the California Alliance for the Mentally Ill, 5*(1), 56–57.

Frese, F. J. (1997a). A consumer/professional's view of ethics in research. In A. E. Shamoo (Ed.), *Ethics in Neurobiological Research with Human Subjects: The Baltimore Conference on Ethics* (pp. 191–194). Amsterdam: Gordon and Breach.

Frese, F. J. (1997b). The mental health consumer's perspective on mandatory treatment. In M. Munetz (Ed.), *Can mandatory treatment be therapeutic? New directions for mental health services*, no. 75 (pp. 17–26). San Francisco, CA: Jossey-Bass.

Frese, F. J. (1998). Advocacy, recovery, and the challenges of consumerism for schizophrenia. In P. A. Buckley (Ed.), *Psychiatric Clinics of North America, 21*(1), 233–249.

Ghaerni, S. N., & Hundert, E. M. (1994). The ethics of research in mental illness. *The Journal of the California Alliance for the Mentally Ill, 5*(1), 47–49.

Goodwin, F. K. (1994). Questions and answers. *The Journal of the California Alliance for the Mentally Ill, 5*(1), 54–47.

Hassner, V. (1994). What is ethical? What is not? Where do you draw the line? *The Journal of the California Alliance for the Mentally Ill, 5*(1), 4–5.

Irving, D. N. (1994). Psychiatric research: Reality check. *The Journal of the California Alliance for the Mentally Ill, 5*(1), 42–44.

Kaminski, R. (1994). The importance of maintaining patient's continued and informed consent. *The Journal of the California Alliance for the Mentally Ill, 5*(1), 57–58.

Kane, J. M. (1994). Enormous ethical challenges. *The Journal of the California Alliance for the Mentally III, 5*(1), 28–29.

Katz, J. (1997). Ethics in neurobiological research with human subjects—final reflections. In A. E. Shamoo (Ed.), *Ethics in neurobiological research with human subjects: The Baltimore conference on ethics* (pp. 329–335). Amsterdam: Gordon and Breach.

Lieberman, J. A., & Sloan, J. (1994). The moral imperatives of medical research in human subjects. *The Journal of the California Alliance for the Mentally Ill, 5*(1), 40–42.

Morgan, D. (1987). Guidelines for the assessment of capacity to consent to medication treatment. Columbus, Ohio: Department of Mental Health.

Munetz, M. (1997). *Can Mandatory Treatment be Therapeutic? New Directions for Mental Health Services*, no. 75. San Francisco, CA: Jossey-Bass.

Munetz, M. R., & Frese, F. J. (2001). Getting ready for recovery: Reconciling Mandatory treatment with the recovery vision. *Psychiatric Rehabilitation Journal, 25*(1), 35–42.

National Bioethics Advisory Commission (1998). *Research involving persons with mental disorders that may affect decisionmaking capacity* (Volume 1). Rockville, Maryland.

Opler, L. A. (1994). Conducting clinical psychiatric research in "non-research" settings. *The Journal of the California Alliance for the Mentally Ill, 5*(1), 30–31.

Pardes, H. (1997). Science and ethics—the search for a balance. In A. E. Shamoo (Ed.), *Ethics in neurobiological research with human subjects: The Baltimore conference on ethics* (pp. 315–324). Amsterdam: Gordon and Breach.

Rubenstein, L. S. (1994). Psychiatric experimentation: The lessons of history. *The Journal of the California Alliance for the Mentally Ill, 5*(1), 22–24.

Schell, B. H. (1994). The ominous shadow of the CIA has imprinted itself on the brain research community. *The Journal of the California Alliance for the Mentally Ill, 5*(1), 38–40.

Shamoo, A. E. (1994). Our responsibilities toward persons with mental illness as human subjects in research. *The Journal of the California Alliance for the Mentally Ill, 5*(1), 14–16.

Shamoo, A. E. (1997). Ethical considerations in medication-free research on the mentally ill. In A. E. Shamoo (Ed.), *Ethics in neurobiological research with human subjects: The Baltimore conference on ethics* (pp. 195–199). Amsterdam: Gordon and Breach.

Shamoo, A. E., & Cassidy, M. M. (1997). Conference summary. In A. E. Shamoo (Ed.), *Ethics in neurobiological research with human subjects: The Baltimore conference on ethics* (pp. 5–26). Amsterdam: Gordon and Breach.

Shore, D., Berg, K., & Mullican, C. (1994). Ethical issues in clinical neurological research. *The Journal of the California Alliance for the Mentally Ill, 5*(1), 61–62.

Smith, M. L. (1994). Power, advocacy and informed consent forms. *The Journal of the California Alliance for the Mentally Ill, 5*(1), 25–27.

Statement of the UCLA Clinical Research Center. (1997). In A. E. Shamoo (Ed.), *Ethics in neurobiological research with human subjects: The Baltimore conference on ethics* (pp. 173–174). Amsterdam: Gordon and Breach.

Straw, P. (1997). Consumer representation on IRBs—How it should be done. In A. E. Shamoo (Ed.), *Ethics in neurobiological research with human subjects: The Baltimore conference on ethics* (pp. 187–190). Amsterdam: Gordon and Breach.

Sunderland, T., & Dukoff, R. (1997). Informed consent with cognitively impaired patients: An NIMH perspective on the durable power of attorney. In A. E. Shamoo

(Ed.), *Ethics in neurobiological research with human subjects: The Baltimore conference on Ethics* (pp. 229–238). Amsterdam: Gordon and Breach.

Weisburd, D. E. (1994). Publisher's note. *The Journal of the California Alliance for the Mentally Ill, 5*(1), 1–2.

Weiss, R. (1998, November 16). Bioethics group divided over research on mentally ill. *The Washington Post*, p. A11.

Weiss, R. (1998, November 18). Research ethics panel urges new regulations to protect mentally ill. *The Washington Post*, p. A11.

Wilson, D. R., Wilson, R., & Tepe, K. (1997). Court-authorized medication for incompetent hospitalized patients. In M. Munetz (Ed.), *Can mandatory treatment be therapeutic? New directions for mental health services*, no. 75 (pp. 73–80). San Francisco, CA: Jossey-Bass.

Wirshing, W. C. (1994). In a perfect world none of this would concern us. *The Journal of the California Alliance for the Mentally Ill, 5*(1), 30.

Witaker, R., & Kong, D. (1998, November 15–18). Doing harm: Research on the mentally ill. *Boston Globe.*

Zarin, D. A. & West, J. (1994). ACORN: APA clinical outcomes research network. *The Journal of the California Alliance for the Mentally Ill, 5*(1), 66–67.

Part VII

# CONFLICTING INTERESTS

CHAPTER 13

# PHARMACEUTICAL INDUSTRY SUPPORT OF PSYCHIATRIC RESEARCH AND EDUCATION: ETHICAL ISSUES AND PROPOSED REMEDIES

CHARLES R. GOLDMAN AND
DAVID L. CUTLER

## INTRODUCTION

Ethical problems associated with conflicts of interest between the needs of the public and the profit motives of industry are having a greater and greater impact on the lives and safety of patients. This is true both in the case of the now infamous managed care industry and also the pharmaceutical industry particularly with respect to it's influence on research and education of physicians. Since the enactment of the Bayh-Dole legislation in 1980, which supported the transfer of research technology from universities to commercial sources, academic research has expanded at a rate of 8.1% annually. From 1980 and 1998 university-generated patents grew from 250 to more than 4800 (Cho et al., 2000). This dramatic increase in academic-industry partnership has produced an impressive array of scientific breakthroughs and new medical treatments. It has also resulted in a remarkable degree of interdependence between academic centers and the pharmaceutical industry. At the end of the 20th century, pharmaceutical companies spent more than $3 billion per year in the United States on clinical drug trials and over $6 billion world wide. Seventy percent of money for clinical drug trials in the U.S. comes from industry (rather than NIH) (Bodenheimer, 2000). At the same time, at least $11 billion is spent each year by pharmaceutical companies in promotion and marketing which amounts to $8000–$13,000 per year on each physician (Wazana, 2000).

Although no hard figures are available, it also appears that industry support of faculty income is increasing, especially through paying faculty to give lectures on topics related to the drugs being marketed (Nemeroff, 1997a; Kirkpatrick, 2000). Academic departments and professional societies increasingly rely on pharmaceutical industry supported speakers to provide education by way of grand rounds, lectures and distance learning conferences. Such a situation creates the potential for conflicts of interest, which could lead to inappropriate clinical decisions and patient suffering.

On the other hand, partnership between academia and pharmaceutical companies may be crucial to the future of medicine. New forms of such collaboration are rapidly evolving, resulting in complex relationships involving such commercial "middle-men" as contract-research organizations (CROs) and site-management organizations (SMOs) which sometimes allow researcher/clinicians to work directly with industry without a significant oversight role for the academic institution. The issues we raise and examine in this chapter reflect the down side to this trend. We will focus on two main areas: the first section examines ethical issues and questions related to the effects of industry funding on research and education activities; the second section discusses proposed remedies. Not addressed in this chapter are problems regarding research fraud, unethical pressure in recruiting research subjects, direct marketing to consumers, and inadequate attention to long term safety and efficacy of medications. In chapter 12, Dr. Frese considers some of these issues.

Although pharmaceutical companies undoubtedly could exercise more self-restraint, we believe that mental health professionals—researchers, clinicians, educators and administrators—must show courageous leadership in facing ethical dilemmas engendered by conflicts of interest in order to assure needed changes. The remedies discussed here are intended to be considered and advocated for by all mental health professionals who are concerned about safe and effective treatment for their clients.

## ETHICAL ISSUES AND QUESTIONS

### RESEARCH

### Does Pharmaceutical Industry Support Lead to Bias in Selection, Design, Implementation, and Reporting of Psychiatric Research?

In a remarkably candid pair of articles on "Striking a balance between clinical medical school departments and the pharmaceutical industry" Charles Nemeroff, M.D., Ph.D., Professor and Chairman of the Department of Psychiatry at Emory University, emphasized the recent growth in dependence of academic departments on pharmaceutical Industry support (Nemeroff, 1997a). Describing "win-win" situations where academic research interests overlap with the profit motive of drug companies, he gave several examples of major academic initiatives funded by the pharmaceutical industry. He also discussed possibilities of abuse in these relationships, stating that the pharmaceutical companies' money "may come with barely perceptible strings attached." In one example, a study comparing a particular drug with placebo and with another drug produced negative results for the sponsoring drug company. Allegedly, researchers were not allowed to report these negative findings on threat of losing future funding. Similar examples, widely reported in the news media and professional journals (Bero & Rennie, 1996; Daly, 1998; Regush, 1998; Schuchman, 1998; Shenk, 1999), described cases where researchers were actively discouraged from reporting negative findings about a drug, including efforts, in some cases, to terminate their employment. Close financial connections resulting in conflicts of interest between academic research departments and the sponsoring companies have been the subject of much media coverage (Wilson & Heath, 2001; Spencer, 2000; Kirkpatrick, 2000; Kauffman & Julien, 2000; Press & Washburn, 2000; Friend, 2000).

Whether or not the interaction between academic researcher and funding organization is seen as advantageous to both, it is obvious that various conflicts of interest can arise. Quoting Thomson (1993), Blumenthal (1996) defines conflict of interest:

> A conflict of interest is a set of conditions in which professional judgment concerning a primary interest ... tends to be unduly influenced by a secondary interest. ... The primary interests are determined by the professional duties of a physician, scholar or teacher. ... They should be the primary consideration in any professional decision. ... In their most general form, the primary

> interests are the health of patients, the integrity of research, and the education of students. ... The secondary interest is usually not illegitimate in itself. ... Only its relative weight in professional decisions is problematic.

Huth (1996, p. 391) lists six parties whose interests may conflict in a clinical research situation: the investigator, the scientific community, the host institution, the commercial sponsors, patients, and the public. Based on Congressional hearings, Huth (p. 391) raises five conflict of interest issues:

1. What are the risks that research might violate scientific standards? What might lead to a flawed study serving a commercial interest at the expense of reliable research?
2. Will industry collaboration shift university research away from basic questions and toward applied and market-oriented research?
3. Will such collaboration reduce the quality of graduate scientific education by minimizing educational objectives and using graduate students as cheap labor?
4. Will industry sponsors limit scientific communication to protect proprietary information?
5. How can an academic institution prevent exploitation of its resources through diversion of faculty effort and laboratory space to the industrial sponsor's interests? How will the federal government prevent diversion of its support for research to proprietary commercial interests at the expense of broader societal interests?

Although these questions do not always have clear answers, the proposed remedies in Section II of this Chapter address most of them.

Other authors writing about ethical concerns raise similar questions. In a series of letters to the editor in 1996–1997 in the Journal of Clinical Psychiatry, several researchers debated ways in which "incentive bias" can influence the results of pharmaceutical studies. Incentive bias occurs when a scholar-researcher is "influenced inadvertently and subtly to render opinions about products or treatments that cleave to the sponsoring company's marketing line" (Gammon, 1996). Jefferson et al. (1997), psychiatrist researchers themselves, further asserted "it would be foolish to deny that incentive bias exists—it influences everyone, including ourselves." Davidson (1986) analysed 107 trials in 5 leading medical journals with regard to outcome and sources of funding and concluded that studies sponsored by pharmaceutical companies were much less likely to favor traditional therapy over new drug treatment. Stelfox et al. (1998), after studying 70 published items, found that authors who had a financial association with manufacturers were much more likely than those who did not to have a favorable published position on the safety of calcium channel antagonists as a treatment for cardiovascular disorders. Only 2 of the 70 articles included in the study disclosed the authors' potential conflicts of interest.

Drug effectiveness studies are extremely complex and there are many opportunities for the design, analysis and reporting to appear to favor a particular drug. Perhaps the biggest source of bias is in the initial decision about which aspects of a treatment to study and which to ignore; e.g., there have been few studies of dependency/withdrawal phenomena in antidepressants, although this is a problem for many of them (Medawar, 1997). Bero and Rennie (1996) cite several examples of bias introduced into research by defining the research question too narrowly. One such example is a study comparing the effects of sertraline, amitriptyline, and placebo on the electrocardiogram, but not on depression. This study was then used to promote sertraline as the best choice to treat depression. In general, Bero and Rennie (1996) conclude that pharmaceutical industry funding of drug studies often results in biases which "make it easier to demonstrate that a new drug is more effective and/or has fewer side effects than the standard therapy."

In randomized, controlled trials, there are opportunities for bias in the selection of the patient population, the blinding or masking procedure, the outcome measures and scores, the comparison treatment, the pre-study medication or "drug wash-out" policy, whether only drug responders are kept in the study, whether active placebos are used, how drop-outs are treated, how adverse reactions are reported, whether placebo responders are kept in, the doses to be compared, the length of the study, the statistical analysis, the way in which results are reported and how the results are summarized (Bero & Rennie, 1996; Brody, 1996, p. 409; Fisher, 1997; Kirsch, 1998; Medawar, 1997; Moore, 1997, 1998; Moncrieff et al., 1998). A discussion of each of these areas is beyond the scope of this chapter.

Even the definition of "research" is open to question, as more industry funding becomes available with little or no requirement that scientific methods be used. For example, often there may be a blurring of the boundaries between research and practice. Clinicians are invited to dinners, vacation resorts, and boat trips and paid fees if they will use specific drugs with their patients in a trial (Kirkpatrick, 2000). Bero and

Rennie (1996) document that studies that are not randomized and well controlled are sometimes misrepresented as if they were.

Friedberg et al. (1999) noted, after searching databases for original English-language research articles of cost or cost-effectiveness analyses of oncology drugs (1988–1998), that pharmaceutical company-sponsored studies were less likely than nonprofit-sponsored studies to report unfavorable qualitative results.

In summary, there is increasing evidence consistent with the hypothesis that pharmaceutical industry support can and does lead to bias in selection, design, implementation, and reporting of psychiatric research. Many authors writing about the subject of industry sponsored research conclude that more research is needed into how potential ethical problems are expressed in real life situations.

## Education

### Does Pharmaceutical Industry Support Lead to Bias in Education of Mental Health Professionals and the General Public?

A major source of information for psychiatric educators, mental health professionals, and the general public is the body of research published in journals. However, what is not published is as important as what is published in forming our impression of effective treatment. As reported by Blumenthal (1996), a 1994 survey of industry-university research centers (IURCs) concluded that 41% had restrictions on their ability to communicate information to the general public, 29% on their communication with faculty at other universities, and 21% on sharing information with faculty in their own institution. Bero and Rennie (1996) list many examples of suppression of negative research findings, including a case where research funds from a pharmaceutical company to a university were revoked when the university adopted guidelines restricting the access of industry representatives to residents.

When such studies are published, sources of research bias are neither consistently explained nor discussed in the journals, nor are peer reviewers routinely asked to comment on possible incentive bias. In fact, the reviewer does not necessarily know the funding source of a research study. The fact that many professional journals are wholly funded by, or greatly dependent on, industry support through direct subsidy and advertising raises additional questions about conflict of interest. Similar conflicts exist in the growing field of Internet based continuing medical education (Kelly, 1998; Silberg et al., 1997).

The peer review process itself is often not defined, or poorly defined, and may vary from a cursory review of an abstract by an editor to a thorough analysis by several impartial experts (Bero & Rennie, 1996). Studies have also shown that articles published in controlled circulation ("throwaway") journals and symposia that are adjuncts to journals have a lower degree of peer review and contain more misleading information than articles in standard professional journals (Bero et al., 1992; Bero & Rennie, 1996; Cho & Bero, 1996). Bero and Rennie (1996) emphasize that literature searches fail to distinguish between articles from peer-reviewed journals vs. those from sources more directly controlled by the pharmaceutical companies.

Even when a published article contains valid and unbiased information, Pitkin et al. (1999) found that the abstract of the article may be biased. In a study of 44 research articles published in six major medical journals, a significant percentage of abstracts was found to be inaccurate, as defined by inconsistency of data or data in the abstract not found in the body of the article. The percent inaccuracy ranged from 18% to 68% depending on the journal. For five of the journals the range was 30–68%.

The preparation of material to be published in a journal can be "ghost written" by employees of the drug company. This may even apply to letters to the editor, as reported by Dr. Nemeroff (1997b):

> An ... example involves Professor B who, having received considerable grant support from Company C, is given a letter prepared by an advertising firm employed by the company, criticizing an article that shows the competitor's pharmaceutical agent in a favorable light. The letter is to be submitted to the journal where the above-cited article is published and it is "authored" by Professor B. A cover letter states that it was prepared by the advertising firm, that Professor B will receive an honorarium for his "work" on the project, and that Professor B should submit the letter directly to the journal from his office.

Increasingly, academic faculty and respected practitioners are being paid by the pharmaceutical industry to speak and write about psychiatric issues (Kirkpatrick, 2000). The conflict of interest issues listed above apply here, too. Nemeroff (1997b) again provides a chilling example:

> Professor A relocates from one major university to another. This individual is then courted by several pharmaceutical company representatives. One company invites the faculty member to present an overview of diagnosis and treatment of a particular disease at a dinner program. After the presentation, the professor is informed that he will never be asked to speak for this company

> again because he did not "sell" their drug during his presentation, said good things about the competitor's drug, and pointed out certain shortcomings of the company's drug. ... Such instances have resulted in faculty members being dropped from pharmaceutical company speaker's bureaus, with corresponding loss of income.

Although many professionals are paid to speak through "unrestricted educational grants" it is often clear that continuation of their speaking career (which may be a significant source of income) may depend on favorable reports from drug company representatives.

In a study examining the content of two courses describing the use of calcium channel blocking agents, each sponsored by the manufacturer of a different agent, clear bias was found in the number of positive mentions of the drug of the sponsor (Shimm, 1996a, p. 326). Another study found that prescribing practice also changed in the direction expected from the course sponsorship (Shimm, 1996a, p. 326). Bero and Rennie (1996) offer many other examples of pharmaceutical industry sponsored publications giving misleading information, which encourage prescribing the sponsor's product.

Being given even minimal compensation for their loyalty may influence professionals (Wazana, 2000). Nemeroff (1997b) gave this example:

> A colleague had to leave town abruptly because of a family illness. Another psychiatrist agreed to cover for him as an attending physician on the inpatient service. The substitute was startled to discover that all of the depressed patients were being treated with the same antidepressant. The psychiatry residents were asked why this was the case. Reluctantly, one said that the representative for that antidepressant bought daily lunches for them and they felt that they owed it to this person, their friend, to prescribe the drug. A similar incident involved a pharmacist who chaired a formulary committee of a major hospital, who admitted that because a particular pharmaceutical representative bought daily lunches for his staff, that company's drug was the preferred agent.

As more academic departments depend on pharmaceutical industry funding, medical faculties feel tremendous pressure to produce results and teach in ways that at least would not alienate the drug companies. A common argument in defense of dependence on pharmaceutical industry support is that if faculty receive funds from several competing drug companies, they do not have to favor any one. This, however, overlooks the fundamental conflict of interest. In general, industry funded lectures promote brand names over generics, emphasize drugs as the most effective treatment, and suggest that drugs should be used as early in an illness as possible, and for as long as possible. These conclusions (as opposed to the ideas that non-drug treatments may be equally effective, that generic drugs work as well as new ones, and that drug treatment is best if it is time limited) translate to many billions of dollars of profit for the companies that sponsor the research and training. Many clever speakers are able to subtly promote one drug in one talk, and another drug to a different audience, without clearly favoring one over the other, but mentioning the sponsor's product by name more frequently than the name of the competing products or in a more favorable light (Shimm et al., 1996a, p. 326). Speakers often express preference for a sub-class of drugs, such as the newer "atypicals" or brand name anticonvulsants (Bero & Rennie, 1996). Another common technique is consistently and subtly to attack the "enemy" drug (whose company does not pay them). Dr. Robert Hsiung, a psychopharmacologist and psychiatrist-educator, agrees that these tendencies exist, further stating that "cause and effect can be hard to sort out. It certainly may be that funding by a drug company affects what a speaker says, but it also may be that it's because of pre-existing biases that a speaker received funding in the first place. A 'pre-biased' speaker doesn't sound as manipulative (or as corrupt). Either way, it's only selected points of view that get voiced, and I suppose that's the important thing" (Hsiung, 1997, personal communication). Jack Freer, M.D., with the Center for Clinical Ethics and Humanities in Health Care, State University of New York at Buffalo, is quoted as stating that the drug companies can seek out academicians who already support their agenda, "then pay them handsomely to travel around speaking. ... It is often quite insidious" (Gianelli, 1998).

In the past, academic presentations on drug studies were relatively boring in that the presenter went to great lengths to highlight the limitations of the studies and to warn that the results could not be generalized beyond the study population. Many qualifications to the findings were presented. Now, the presentations and publications often are streamlined with high tech photos and slides and the findings are exaggerated and overly generalized (Bero & Rennie, 1996). This is more attractive to students and audiences in general, but potentially misleading. Increasingly, the source of funding can be surmised from the presentation. Robert Bollinger, Ph.D., director of CME for Wayne State University School of Medicine, is reported as believing that, even when all the involved parties adhere to every relevant guideline, the drug companies still drive the curriculums (Gianelli, 1998).

Well-known and respected psychiatrists, social workers, nurses, patient educators, and other professionals are accepting pharmaceutical industry funding to support their writing, speaking, consultation and patient education activities. They may be addressing topics unrelated to the sponsor's product. Many are even providing these services without charge in the belief that the materials will help consumers and contribute to the profession. These people are generally not biased and may not even use the sponsor's product, although they tend to support the use of medications in general. When specific drug promoting talks are presented (or written in industry sponsored publications), they can be "sandwiched" between presentations by the non-biased (or less biased) clinicians. The most clever, and probably most effective, drug company marketing is subtle and disguised by its close association with neutral clinical material.

In summary, industry support for research, speakers, publications, and internet sites may well lead to bias in education of mental health professionals and the general public.

## WHAT CAN WE DO?

### Remedies for Conflicts of Interest in Research

Many authors writing about the ethics involved in the relationship between clinical research and the pharmaceutical industry have suggested safeguards and remedies for the problem areas described above. Dr. Nemeroff, after highlighting some of the abuses that occur, urged practitioners to be mindful of the temptations and to adhere to basic ethical behavior as spelled out in existing guidelines (Nemeroff, 1997b).

There is considerable evidence, however, that existing guidelines are inadequate. In a critique of the FDA, for example, Moore (1997) asserted that this regulatory body is far too friendly with the pharmaceutical industry to be objective. He also documented the inadequate funding and weak mandate for the FDA to monitor the safety and efficacy of medications once they are initially approved and to address long term use of medications (Moore, 1998). A USA Today study of the FDA drug review process (Cauchon, 2000) found that 54% of supposedly independent experts had a direct financial interest in the drug or topic they were asked to evaluate. Although Federal law prohibits the FDA from using experts with financial conflict of interest, the FDA waived the restriction more than 800 times between January, 1998, and June, 2000. Analysis of 159 advisory committee meetings found:

- at 92% of meetings, at least one member had financial conflict of interest;

- at 55% of meetings, half or more FDA advisors had financial conflict of interest;
- conflicts were most frequent at the 57 meetings when broader issues were discussed (92% of members had conflicts);
- at the 102 meetings dealing with the fate of a specific drug, 33% had a conflict.

Several authors have proposed reforms for the FDA (e.g., Bero & Rennie, 1996; Angell, 2000b).

In 1996 the Public Health Service issued guidelines requiring all academic researchers to report to their schools if they have received payments of more than $10,000 from a company or if they hold at least 5% of its stock. At most universities this information is kept private (Press & Washburn, 2000). Existing Federal rules regarding conflicts of interest in research leave details of policy enforcement to institutions, where there is "danger of lax enforcement, abuse and scandal" (Kodish et al., 1996). In an extensive review of the structure, composition and function of local academic Institutional Review Boards (IRBs), Francis (1996) demonstrated that there are many ambiguities and areas where such Boards may themselves be subject to conflicts of interest. Further, research done on IRBs has revealed that few provide training for members regarding issues related to conflicts of interest in research and, in general, such Boards do not adequately address the issues raised in this Chapter (Brown, 1998).

Kodish et al. (1996) pointed out that existing Federal guidelines focus exclusively on financial conflicts and ignore "other unwanted consequences of university–industry relationships, for example, the growth of secrecy within university science" and the adverse effects of such secrecy on the disinterested pursuit of objective scientific truth and on the training of future scientists. However, other authors have emphasized that financial conflicts pose a much different and more urgent problem than non-financial conflicts (such as pressure on faculty to publish positive findings). Some have proposed that conflict of interest policies for research that does not require human subjects be considered separately from patient-oriented research. For basic research, Martin & Kasper (2000) believe it is permissible, with full disclosure to the institution, for investigators to receive financial support from companies from which they receive consulting fees. For clinical research they propose much more strict guidelines and oversight.

Cho et al. (2000) studied policies on faculty conflicts of interest at major US universities and found that fifty-five percent of policies

($n = 49$) required disclosures from all faculty while 45% ($n = 40$) required them only from principal investigators or those conducting research. Nineteen percent of policies ($n = 17$) specified limits on faculty financial interests in corporate sponsors of research, 12% ($n = 11$) specified limits on permissible delays in publication, and 4% ($n = 4$) prohibited student involvement in work sponsored by a company in which the faculty mentor had a financial interest. They concluded that most policies on conflict of interest lack specificity about the kinds of relationships with industry that are permitted or prohibited.

In a thorough review of existing guidelines for addressing conflicts of interest in industry-funded clinical research, Huth (1996) summarized guidelines from the American College of Cardiology, American College of Physicians, American Medical Association, American Psychological Association, Royal College of Physicians, US National Academy of Sciences, The Institute of Medicine, National Science Foundation, Public Health Service, National Council on Bioethics in Human Research, and various medical journals. Although favoring the guidelines of the National Council on Bioethics in Human Research, Huth found all of the existing guidelines lacking. His extensive recommendations included:

- Expectations of research sponsors should be clearly spelled out up front, including all financial arrangements and procedures for terminating study and reporting results.
- Payment should go to institutions, and not to individual researchers.
- Institutional review boards should be specifically trained to detect potential incentive bias and should review the above expectations and should have a crucial role in the entire process, including negotiations of compensation.
- On completion of research, all findings should be written up, including those unfavorable to sponsor.
- Investigators should be free to report all findings.

Many authors have praised the strict Harvard Medical School policy on faculty relationships with commercial entities: Policy on conflicts of interest and commitment (adopted by the Harvard Medical Center May 16, 1990; amendments adopted December 13, 1993, and December 18, 1995. Available at *http://www.hms.harvard.edu/integrity/*). This policy was reaffirmed in 2000 after an extensive review in which many faculty argued that the policy should be relaxed in order to counter the growing trend for faculty to leave the university in order to form business relationships with industry. The Harvard policy represents a model for a prescriptive approach to the problem. Another organization which has adopted a strict policy is the American Society of Gene Therapy, whose members must either refrain from participation in a study sponsored by a company in which they have a financial interest or, alternatively, give up that interest.

Not many universities are likely to withstand faculty and industry pressures that would result from such strict policies, however, and a more flexible and adaptable approach has been recommended. The Harvard Medical School sponsored a meeting from which emerged a Consensus Statement on Conflict of Interest Policies for Academic Institutions (personal communication, 2001). This statement represents a major step forward in developing uniformity while also allowing for individual differences among research centers. The consensus statement is included in this chapter in its entirety as *APPENDIX A*.

Although detailed guidelines such as Huth's and those in *Appendix A* are needed, it is unlikely they could be made mandatory, and be adequately enforced, in a free market culture. In addition to suggesting guidelines, several authors have proposed formation of a broadly constituted study panel or task force to make recommendations for future action. For example, Tenery (2000) recommended that a task force of representatives of the medical profession and pharmaceutical industry be convened to develop standards of conduct addressing potential conflict of interest and disclosure. And Angell (2000b) proposed an independent national advisory panel to study the pharmaceutical industry's practices thoroughly and then make recommendations.

Others have gone further to recommend the formation of a permanent agency or committee. Bero and Rennie (1996) proposed formation of a center for the assessment of pharmaceutical effectiveness that would investigate relative effectiveness and would be funded by subscription fees on payers, contributions by payers to research specific questions, and a tax on pharmaceutical products. Sheldon Krimsky, a Tufts University professor, suggested creating a government fund for clinical testing of drugs, to put some distance between the companies and the researchers (Kauffman & Julien, 2000). Dr. Raymond Woosley, chairman of pharmacology at Georgetown University, proposed several national "centers for education and research in therapeutics" that would evaluate and compare drugs already on the market. After three years of lobbying, Congress approved the idea. But while Woosley envisioned a $75 million annual budget, Congress appropriated just $2.5 million (Kauffman & Julien, 2000).

We (Goldman and Cutler) suggest that a voluntary Consortium of academic departments, ethics institutes, consumer protection organizations, governmental agencies and pharmaceutical companies be formed to promulgate strict guidelines such as those described. Clinical research programs could then apply for membership in this Consortium, much as clinical programs now apply for accreditation in order to meet the voluntary guidelines (standards) of JCAHO or CARF. Membership in the Consortium would involve ongoing monitoring and record keeping to document adherence to the ethical guidelines. Research reports produced by member organizations could carry the Consortium's certification and thus be accorded the extra respect they would deserve. Certified organizations would be in an excellent competitive position to attract additional funding. Pharmaceutical companies belonging to the Consortium would also have more credibility and this could be marketed so that it would translate into profits. As the work of the Consortium became known, more and more companies and programs would become interested in adhering to the guidelines, or standards, for ethical research.

Finally, Blumenthal (1996) recommends that some degree of financial conflict of interest be tolerated, but only where "the potential financial gain to academic investigators or institutions may be so small that the likelihood of harm to patients resulting from conflict of interest falls below a threshold justifying regulation. ... For example, investigators may be permitted to receive occasional, modest, non-recurring consulting payments from companies sponsoring their clinical research or to own small numbers of shares in publicly traded companies from which they receive clinical research support. ... Institutions may be allowed to own modest positions in publicly traded companies funding clinical AIRs conducted by their faculties."

Blumenthal calls for more research into the effect of industry support on the integrity of research where there are no direct financial gains except the research support itself and a wish that it be continued. He worries about researchers who view science as "simply another valuable human activity that may be executed well or poorly. ... [To these individuals,] compromising the integrity of research may reduce its efficiency and undermine its public support, but ill effects need to be balanced against any gains that may be realized from the activities that cause these side effects." Rather, he prefers researchers who "find their responsibility to protect the integrity of the research every bit as compelling as protecting the welfare of human subjects of research or trainees. Science, from this perspective, is a priestly calling."

## Remedies for Conflicts of Interest in Education

As documented earlier in this Chapter, there are many ways in which the quality of medical information can be compromised. We need clear standards "by which to judge the quality of editorial content, to differentiate author from shill, editorial from advertising, education from promotion, evidence from opinion, science from hype" (Silberg et al., 1997).

Existing guidelines addressing conflicts of interest in education funded by the pharmaceutical industry are brief and somewhat vague. For example, the AMA's code of ethics on gifts to physicians from industry state that "when companies underwrite medical conferences or lectures other than their own, responsibility for and control over the selection of content, faculty, educational methods, and materials should belong to the organizers of the conferences or lectures" (Council on Ethical and Judicial Affairs, 1991).

The American College of Physicians position paper (American College of Physicians, 1990) proposes a more stringent set of guidelines:

1. Gifts, hospitality, or subsidies offered to physicians by the pharmaceutical industry ought not to be accepted if acceptance might influence or appear to others to influence the objectivity of clinical judgment.
2. Independent institutional and organizational continuing education providers that accept industry-supported programs should develop and enforce explicit policies to maintain complete control of program content.
3. Professional societies should develop and promulgate guidelines that discourage excessive industry-supported gifts, amenities, and hospitality to physicians at meetings.
4. Physicians who participate in practice-based trials of pharmaceuticals should conduct their activities in accord with basic precepts of accepted scientific methodology.

The American College of Physicians (1990) guidelines also state that "a useful criterion in determining acceptable activities and relationships is: Would you be willing to have these arrangements generally known?"

Davidoff (1997), the editor of the Annals of Internal Medicine, argues that "given the enormous variability of individual-industry links ... it does not seem appropriate to assume a priori that a link between an individual author and industry automatically creates a meaningful conflict of interest; this is in contrast to the situation of industry support for research projects". His journal requires authors, in a cover letter, to

> disclose any financial interests, direct or indirect (dual commitment), that might affect the conduct or reporting of the work they have submitted. If the authors are uncertain about what might be considered a dual commitment, they should err on the side of full disclosure. Information about dual commitment may be made available to reviewers. If, in the Editors' judgment, a dual commitment represents a potential conflict of interest, information concerning the relationship may be published at the Editors' discretion; authors will be informed of the decision before publication.

The policies of the American Journal of Psychiatry similarly allow for editorial discretion, in consultation with the authors, as to whether commercial interests will be disclosed in the published article (Editor, 1998).

In a brief review of the topic of financial disclosure in publications, Krimsky and Rothenburg acknowledge a range of policies, from downplaying the importance of disclosure and making it voluntary, on the one hand, to requiring full disclosure of any potential conflict, on the other. They state that "the International Committee of Medical Journal Editors (ICMJE) has identified 'financial relationships with industry (for example, employment, consultancies, stock ownership, honoraria, expert testimony), either directly or through immediate family,' as the most important conflicts of interest." They recommend that journal editors "should begin to take seriously the implementation of disclosure policies in response to the escalation of financial interests of authors. ... Journals should be specific in their instructions to authors on the types of financial associations related to their submission and the form of communication (original research, letters, book reviews, and scientific review articles) that warrant disclosure. We also believe that the scientific community and the public will be better served by the open publication of financial disclosures for readers and reviewers to evaluate."

Of course, even the most thorough disclosure policy is useless if not strictly adhered to by the editors. In a dramatic editorial in the New England Journal of Medicine, the editors admitted that the journal had failed to follow its own conflict-of-interest policy in Drug Therapy articles (18 instances out of 42 review articles from 1997 to 2000) (Angell et al., 2000).

In another important area, Bero and Rennie (1996) make three recommendations to ensure that data from drug studies are published:

> First, pharmaceutical company funders should in no way restrict publication by their funded researchers. [These companies should]

> commit themselves to submitting for publication every study that is undertaken, regardless of the findings. This commitment should also extend to leaving the editing ... and the selection of data for publication to the investigators, not the funding company. ...
>
> Second, regulatory authorities should require publication of all data that have been submitted as part of the drug approval process. ... Third, [registries of clinical trials should be maintained]. Such registries would facilitate tracking down negative drug studies that are not published in the medical literature.

We agree with the above recommendations, and offer the following further suggestions:

1. Journal editors, reviewers, and sponsors of educational events should review offerings that are industry funded with incentive bias in mind.
2. Any educational presentations or publications with pharmaceutical company funding should contain warnings that results may be biased.
3. Disclosures by authors should be made more conspicuous, should include the relative amount and type of money involved (e.g., honoraria and expenses), and warnings that highlight potential conflicts of interest.
4. The exact role of drug company employees should be clearly specified (e.g., editorial or authorship roles, ghostwriting).
5. All pharmaceutical company sponsored articles and presentations, whether original research or reviews of the literature, should highlight the study or review limitations, offer alternative explanations of the results and specify all possible limitations to generalization of results.
6. Medical and other professional students should receive training in critical appraisal of research, including issues raised in this chapter. An expanding list of resources exists on the internet to help individuals evaluate the quality of medical information (e.g., see *http://www.dr-bob.org/quality.html;http://www.cochrane.org*). Guidelines and checklists for this purpose are sponsored by such organizations as the American Telemedicine Association *(http://www.atmeda. org/news/ 072899.html)* and the Health Information Technology Institute of Mitretek Systems *(http://hiti-web.mitretek.org/docs/policy.html)*.
7. There should be professional journals that are independent of drug company support (like Consumer Reports). There are a few now (e.g., Community Mental Health Journal, Schizophrenia

Bulletin) which could become even more proactive in encouraging articles investigating use of older medications and non-medication treatments. Some very useful internet sites are non-industry supported and could similarly be expanded.

8. Professional journals and web sites that do accept advertising or pharmaceutical company funding should develop mechanisms to review advertising claims and insulate advertising departments from editorial processes. The American Medical Association has published Guidelines for medical and health information sites on the internet (Winker, 2000).
9. Non-pharmaceutical support for education, including journals and internet sites, is needed. It has been suggested that such support be sought from the managed care industry, other third-party payers, the federal government, foundations, subscribers, a tax on pharmaceutical products, higher tuition, and physicians themselves, who "are in the top 1% of earners in the U.S." (Psychiatric News, 7/3/98).

Finally, we agree with Angell's (2000a) assertion that much more needs to be done to protect students from the now pervasive influence of industry gifts and favors. She says "teaching hospitals should forbid drug-company representatives from coming into the hospital to promote their wares and offer gifts to students and house officers. House officers should buy their own pizza, and hospitals should pay them enough to do so. ... Similarly, academic medical centers should be wary of partnerships in which they make available their precious resources of talent and prestige to carry out research that serves primarily the interests of the companies. That is ultimately a Faustian bargain."

## CONCLUSION

The integrity of, and public respect for, clinical research and professional education is endangered by inadequate attention to possible conflicts of interest and conflicts of commitment. All funding sources carry some risk of encouraging "incentive bias," and no reasonable critics have proposed doing away with pharmaceutical company sponsorship of research and training. Several have urged the pharmaceutical industry to assume a more responsible role in addressing these ethical concerns, even arguing that this industry "should be held to owe a duty of utmost good faith and concern ('fiduciary duty') for consumers' safety and well being" (Shimm et al., 1996b). However, the larger burden of responsibility falls on professionals who clearly have a fiduciary duty to place the interests of patients, research subjects, and science above personal welfare.

Several remedies have been proposed for situations that create potential for conflicts of interest in pharmaceutical support of psychiatric research and education. As Press and Washburn (2000) concluded, "universities could do more to make the case for preserving public support for higher education while refusing to tailor either the research agenda or the curriculum to the needs of industry. ... The ultimate criterion of the place of higher learning in America will be the extent to which it is esteemed not as a necessary instrument of external ends, but as an end in itself."

## REFERENCES


American College of Physicians (1990). Physicians and the pharmaceutical industry. *Annals of Internal Medicine, 112*: 624–624.

Angell, M. (2000a). Is academic medicine for sale? *N. Engl. J. Med., May 18; 342*(20), 1516–8.

Angell, M. (2000b). The pharmaceutical industry—to whom is it accountable? *N. Engl. J. Med., Jun 22; 342*(25), 1902–1904.

Angell, M., Utiger R. D., Wood A. J. (2000). Disclosure of authors' conflicts of interest: A follow-up. *N. Engl. J., Med. Feb 24; 342*(8), 586–587.

Bero, L. A., & Rennie, D. (1996). Influences on the quality of published drug studies. *International Journal of Technology Assessment in Health Care, 12*(2), 209–237.

Bero, L. A., Galbraith, A., & Rennie, D. (1992). The publication of sponsored symposia in medical journals. *N. Engl. J. Med., 327*(16), 1135–1140.

Blumenthal, D. (1996). Ethics issues in academic-industry relationships in the life sciences: the continuing debate. *Acad. Med., 71*, 1291–1296.

Bodenheimer, T. (2000). Uneasy alliance—clinical investigators and the pharmaceutical industry. *N. Engl. J. Med., May 18; 342*(20), 1539–1544.

Brody, B. A. (1996). Conflicts of interest and the validity of clinical trials. In Jr. Spece. R. G., D. S. Shimm, & A. E. Buchanan (Eds.), *Conflicts of interest in clinical practice and research.* (pp. 407–417).

Brown, J. B. (1998) Institutiuonal Review Boards: Their Role in Reviewing Approved Research. Department of Health and Human Services, Office of Inspector General OEI-01-97-00190.

Cauchon, D. (2000). USA TODAY study of FDA drug review process. *USA Today* (9/25/00).

Cho, M. K., & Bero, L. A. (1996). The quality of drug studies published in symposium proceedings. *Ann. Intern. Med., 124(5)*, 485–489.

Cho et al. (2000). Policies on faculty conflicts of interest at US universities. *JAMA Nov 1; 284*(17), 2203–2208.

Council on Ethical and Judicial Affairs (1991). Gifts to physicians from industry. *JAMA, 265*(4), 501–501.

Daly, R. (1998). Doctors divided on drug study. *The Toronto Star*, 8/14/1998 (*www.thestar.com/back_issues ... 14/news/980814NEW09_CI-DRUG14.html*)

Davidoff, F. (1997). Where's the bias? *Ann. Intern. Med., 126,* 986–988.

Davidson R. A. (1986). Source of funding and outcome of clinical trials. *J. Gen. Intern. Med., 1,* 155–158.

Editor. (1998). Information for authors. *Am. J. Psychiatry, 155*(8), A39.

Fisher, S., & Greenberg, R. P. (1997). What are we to conclude about psychoactive drugs? Scanning the major findings. In S. Fisher, & R. P. Greenberg (Eds.), *From placebo to panacea: Putting psychiatric drugs to the test.* (pp. 359–384). New York: John Wiley & Sons, Inc.

Francis, L. (1996). IRBs and conflicts of interest. In Jr. Spece R. G., D. S. Shimm, & A. E. Buchanan (Eds.), *Conflicts of interest in clinical practice and research.* (pp. 418–436).

Frankel, M. S. (1996). Perception, reality, and the political context of conflict of interest in university-industry relationships. *Acad. Med., 71,* 1297–1304.

Friedberg, M. et al. (1999). Evaluation of conflict of interest in economic analyses of new drugs used in oncology. *JAMA Vol. 282 No. 15,* 1453–1457.

Friend, T. (2000). Conflicting interests compromise, *USA TODAY* (Feb. 22, 2000). *http://www.usatoday.com/life/health/genetics/therapy/lhgth025.htm*

Gammon, G. D. (1996). Incentive bias? *J. Clin. Psychiatry, 57,* 265.

Gianelli, D. M. (1998). Revisiting the ethics of industry gifts. *American Medical News,* 8/24/1998, 7–9.

Hautzinger, M. (1998). Should medicine thumb nose at drug company support for education? *Psychiatric News, 7/3/1998, 23,* 14–28.

Huth, E. J. (1996). conflicts of interest in industry-funded clinical research. In Jr. Spece. R. G., D. S. Shimm, & A. E. Buchanan (Eds.), *Conflicts of interest in clinical practice and research.* (pp. 389–406). New York: Oxford University Press.

Jefferson, J. W., Greist, J. H., & Katzelnick, D. J. (1997). Continuing dialogue on incentive bias. *J. Clin. Psychiatry, 58,* 450–452.

Kauffman, M., & Julien, A. (2000). Surge in corporate cash taints integrity of academic science. The *Hartford Courant* April 9, 2000 *http://www.ctnow.com/scripts/editorial.dll? render=y&eetype=Article&eeid=2037340&ck=&uh=320421005,2,&ver=2.5.*

Kelly, C. K. (1998). The hype, and concerns, about CME on the Internet. *ACP Observer, February, http://www.acponline.org/journals/news/feb98/cmehype.htm.*

Kirkpatrick, D. D. (2000). Cover story—Inside the happiness business. *New York Magazine* May 15, 2000 *http://www.newyorkmag.com/page.cfm? page_id=3122&position=1.*

Kirsch, I., & Sapirstein, G. (1998). Listening to Prozac but hearing placebo: a meta-analysis of antidepressant medication. *Prevention & Treatment, 1,* 1–15.

Kodish, E., Murray, T., & Whitehouse, P. (1996). Conflict of interest in university-industry research relationships: realities, politics, and values. *Acad. Med., 71,* 1287–1290.

Korn, D. (2000). Conflicts of interest in biomedical research. *JAMA Nov 1; 284*(17), 2234–7.

Krimsky, S., & Rothenberg, L. S. (1998). Financial interest and its disclosure in scientific publications. *JAMA, 280,* 225–226.

Kurt, T. L. (1996). Regulation of government scientists' conflicts of interest. In Jr. Spece R. G., D. S. Shimm, & A. E. Buchanan (Eds.), *Conflicts of interest in clinical practice and research.* (pp. 377–388). New York: Oxford University Press.

Lane, R. M. (1996). Incentive bias? *J. Clin. Psychiatry, 57,* 265–267.

Medawar, C. (1997). The antidepressant web: marketing depressiona and making medicines work. *International Journal of Risk & Safety in Medicine, 10,* 75–126.

Moncrieff, J., Wessely, S., & Hardy, R. (1998). Meta-analysis of trials comparing antidepressants with active placebos. *British Journal of Psychiatry, 172,* 227–231.

Moore, T. J. (1997). Hard to swallow. *Washingtonian, December*, 68–145.
Moore, T. J. (1998). Fixing the sytem. In T. J. Moore (Ed.), *Prescription for Disaster*. (pp. 172–188). New York: Simon & Schuster.
Nemeroff, C. B. (1997b). Striking a balance with the pharmaceutical industry: part II. *CNS Spectrums, 2*, 68.
Nemeroff, C. B. (1997a). Striking a balance between clinical medical school departments and the industry. *CNS Spectrums, 2*, 65–66.
Press, E., & Washburn, J. (2000). The kept university. *The Atlantic Monthly, March, 2000*, pp. 39–54 *http://www.theatlantic.com/issues/2000/03/index.htm.*
Pitkin, R. M. et al. (1999). Accuracy of data in abstracts of published research articles. *JAMA Mar 24–31; 281*(12), 1110–1.
Regush, N. (1998). Don't tell, don't ask. ABC News. Health and Medical Issues in the News. (October 20, 1998) *http://abcnews.go.com/sections/living/SecondOpinion/secondopinion3.html.*
Rodwin, M. A. (1993). *Medicine, Money, and Morals*. New York NY: Oxford University Press.
Schuchman, M. (1998). Secrecy in science. *Ann. Intern. Med., 129*, 341–344.
Shenk, D. (1999). Money + Science = Ethics Problems on Campus. Feature Story in *The Nation* March 22, 1999 *http://past.thenation.com/cgi-bin/framizer.cgi?url=http://past.thenation.com/issue/990322/0322shenk.shtml.*
Shimm, D. S., Spece, R. G., Jr., & DiGregorio, M. B. (1996a). Conflicts of interest in relationships between physicians and the pharmaceutical industry. In Jr. Spece. R. G., D. S. Shimm, & A. E. Buchanan (Eds.), *Conflicts of interest in clinical practice and research*. (pp. 321–357). New York: Oxford University Press.
Shimm, D. S., & Spece, R. G., Jr. (1996b). An introduction to conflicts of interest in clinical research. In Jr. Spece R. G., D. S. Shimm, & A. E. Buchanan (Eds.), *Conflicts of interest in clinical practice and research*. (pp. 361–376). New York: Oxford University Press.
Silberg, W. M., Lundberg, G. D., & Musacchio, R. A. (1997). Assessing, controlling, and assuring the quality of medical information on the internet. *JAMA, 277*, 1244–1245.
Spencer, S. (2000). Drug Money: Medical Trials Run Without Real Doctors Elaborate Fraud Scheme Lined Researchers' Pockets Patients Worsened; Little Oversight Provided. CBS News "48 Hours" (7/31/2000, Augusta, Georgia) *http://wfor.cbsnow.com/now/story/0,1597,220233-412,00.shtml*
Stelfox et al. (1998). Conflict of interest in the debate over calcium-channel antagonists. *N. Engl. J. Med. Jan 8; 338*(2), 101–106.
Tenery, R. M. (2000). Interactions between physicians and the health care technology industry. *JAMA, January 19– vol 283*: 391–393.
Thomson, D. (1993). Understanding financial conflicts of interest. *N. Engl. J. Med., 329*, 573.
Wazana, A. (2000). Physicians and the pharmaceutical industry: is a gift ever just a gift? *JAMA Jan 19; 283*(3): 373–380.
Wilson, D. & Heath, D. (2001). Uninformed consent: a five-part Seattle Times investigative series. *Seattle Times 3/19/01 http://seattletimes.nwsource.com/uninformed_consent/.*
Winker et al. (2000). Guidelines for medical and health information sites on the internet: principles governing AMA web sites. American Medical Association. *JAMA Mar 22–29; 283*(12), 1600–1606.

# APPENDIX A

From: Joseph B. Martin Ph.D., M.D.
Caroline Shields Walker Professor of Neurobiology and Clinical Neuroscience
Dean of the Faculty of Medicine
Harvard Medical School
Received 3/5/01
Note: The Statement came out of a meeting held in Washington, D.C., on 11/27/00 and 11/28/00 called by Harvard Medical School Dean Joseph Martin. The attendees were Primarily Deans of the largest NIH recipient medical schools but included a few others with an interest in this area. There was not a specific author and therefore it is called the Consensus Statement. (http://www.hms.harvard.edu/news/releases/020801conflict.html)

## Consensus Statement on Conflict of Interest Policies for Academic Institutions

### Statement of Purpose (01/23/01)

The protection of human research subjects and the integrity of biomedical research are of paramount importance to American medical schools, teaching hospitals, and research institutes. Industrial collaborations are essential if patients are to benefit from the translation of biomedical research into clinical practice. However, the potential financial conflicts of interest that may arise from these relationships require that we have consistent and adequate standards for managing such conflicts. We are therefore proposing a set of principles and guidelines to be used by American medical schools, teaching hospitals, and research institutes as they review and refine their individual institutional policies. In proposing these principles and guidelines we note that we are addressing potential conflicts of interest on the part of individuals and not of institutions. Issues of institutional conflict of interest are also important ones which merit separate and careful review and consideration.

## Proposed Guidelines on Policy Issues

- Every medical school and research institution should have a written policy on financial interests related to research.
- The policy should apply to individuals who are directly involved in the conduct, design, or review of research including faculty, trainees, students and administrators.

- The policy should include both a statement of general principles and a clear delineation of the activities and the levels and kinds of financial interests related to research that are and are not permissible, and/or that require review and approval. The policy should specifically address the special circumstances surrounding research involving human subjects. Individuals involved in the conduct, design, or reporting of research involving human subjects should not have more than a clearly defined minimal personal financial interest in a company that sponsors the research or owns the technology being studied.
- Financial interests covered by the policy should include fees, honoraria, gifts and other emoluments for consulting or lecturing; equity interests including stock options and expectations of receiving equity interests; and directorships, executive roles, and other special relationships with companies having the potential for personal material gain.
- The policy should stipulate whose financial interests, in addition to those of an individual involved in the research, could pose a conflict of interest for that individual.
- All key terms in the policy, such as "family" and "financial interests," should be clearly defined.
- Any financial interests deemed by the institution to be allowable, such as equity interests in mutual funds, should be clearly delineated in the policy.
- The policy should clearly state the procedures to be followed in disclosing financial interests, reviewing disclosure forms, implementing the policy, appealing decisions concerning the policy, and sanctioning non-compliance with the policy.
- The policy should clearly define the range of possible sanctions for non-compliance with the provisions of the policy, up to and including dismissal, and reference the procedures to be followed in the sanctioning process.
- There should be coordination within the various offices of the institution dealing with research and conflict of interest, including committees on human subjects protection, offices of technology transfer, and other related functions.

## Proposed Guidelines on Disclosure

- Faculty, trainees, students, and staff who participate in research should periodically and prospectively disclose all related financial

interests; interim updates should be required whenever situations change.
- Disclosure of related financial interests should be made to specifically designated institutional offices and to the research funder. In the case of funding by a federal agency, disclosure should be made in conformance with federal requirements.
- Faculty, trainees, students, and staff who participate in clinical research should disclose related financial interests to institutional review boards (IRBs). Each IRB should have responsibility for ensuring that patients are informed of such relationships as the IRB determines is appropriate
- Faculty, trainees, students, and staff should disclose all related financial interests in any publications and presentations, including presentations made both within and without the institution.
- Biomedical science journals should be encouraged to require and to publish the disclosure of related financial interests.

## Guidelines on Implementation and Review

- Disclosure is an essential component of managing potential financial conflicts of interest; consequently, every institution must have an explicit policy on disclosure, both to the institution and to outside entities.
- A mechanism must exist to assure dissemination of the policy to faculty, staff, and students, and to provide appropriate education and training in the policy.
- Faculty and research staff should formally acknowledge that they have read and understand the policy.
- There should be requirements for regular periodic reporting as well as interim updates utilizing a reporting disclosure form.
- Disclosure should be made to multiple levels within each institution including the Dean, CEO, or the equivalent individual, who has ultimate responsibility for monitoring the activities of the faculty, staff and students, and to the department chair(s).
- Each institution should have an advisory policy oversight committee which has broad representation of faculty, administrative staff, and possibly lay representatives; the committee should be charged with:
  - Providing oversight of the policy
  - Reviewing cases that are brought before the committee

- Recommending monitoring procedures for exceptional cases when appropriate.

- Monitoring policies and procedures should be prospectively defined
- The oversight committee should be advisory the Dean, CEO, or equivalent individual who may appoint an ad hoc monitoring committee, the composition of which should be by case-specific issues, when appropriate. Final authority for specifying monitoring in a specific circumstance should be the responsibility of the Dean, CEO or equivalent individual.
- Overall institutional compliance with the policies should be monitored using the institution's internal audit mechanisms.
- Conflicts between faculty should be resolved by the advisory oversight committee with recommendations to the Dean, CEO or equivalent individual who has the ultimate authority to define the terms of a final resolution.

# INDEX